M000313488

THE PROPTECH GUIDE

EVERYTHING YOU NEED TO KNOW ABOUT
THE FUTURE OF REAL ESTATE

LAWIN CHANDRA

Copyright © Lawin Chandra

All rights reserved.

No part of this book may be reproduced, distributed, or transmitted in any form or by any means, including photocopying, recording, or other electronic or mechanical methods, without the prior written permission of the publisher, except in the case of brief quotations embodied in reviews and certain other non-commercial uses permitted by copyright law.

Line Editing:	Douglas Williams
Copy Editing:	Jeff Rudolph
Interior Layout:	Sandeep Likhar
Cover Design:	CrowzArt Design

ISBN:

978-1-9164796-0-9 (e-book)
978-1-9164796-1-6 (paperback)
978-1-9164796-2-3 (hardcover)
978-1-9164796-3-0 (audiobook)

Published by

PropTech Asset Management LTD
17 Hanover Square
London
W1S 1BN
United Kingdom
www.PropTechAM.com

Table of Contents

1

Introduction to PropTech

"When the wind of change
blows, some people build
walls, others build windmills."

–Ancient Chinese proverb

Real Estate Problems

The real estate market used to have a solid foundation and most considered it to be stable. A number of investment gurus considered an investment into a commercial or residential property as a safe bet in order to hedge against a volatile investment product. The stable rental income on a monthly basis gave a feeling of high security and stability, where future returns could be easily forecast and accurately predicted.

Today, however, real estate seems to have more problems than solutions, having caused a bubble and the ensuing financial crisis in 2008, and is now seemingly outdated, as old ways of doing things endure while technological advancements skirt the industry, to the industry's detriment. Before the financial crisis, participants correctly assumed the underlying asset consisted of brick and mortar, but only a few participants questioned whether the tenant or occupier of the underlying asset was able to pay the rent, or able to repay the monthly mortgage. Researching the creditworthiness of a tenant or a borrower was and still is a rather difficult adventure to undertake. So many different intermediaries are involved in every aspect of a real estate activity, and little transparency is given.

There seem to be too many market players involved in a transaction, each benefiting from a lack of transparency that produces massive delays and very high costs. Just think about how many unnecessary market players are involved in such a transaction. There is a seller, his broker, the buyer, the buyer's broker, the buyer's solicitor, the seller's solicitor, a notary, a mortgage broker, a bank, and the surveyor. And all these participants want to earn money by playing on their specialist knowledge, which they don't want to fully disclose to other players.

In most cases, these participants use software to become more efficient, but the software is proprietary and rarely compatible with what is being used by other players.

Real Estate & Technology Expertise

Before I get into my discussion of how to improve real estate operations, let me step back and provide a little insight on who I am and how I have come across my expertise.

For the last 15 years, I have been working in different industries, countries, and positions, where I was always questioning and improving the status quo of a situation.

One of my first jobs was to work for the large blue-chip company Electronic Arts, in Spain, which was and is still publishing the best video games of all time. The technology used for their software ten years ago is now translating into virtual reality and augmented reality used in real estate through off-plan project development presentations. I was involved in the development and management of the software globally.

The people I met broadened my network in the high-tech environment and strongly influenced my view toward the utilization of technology in other industries, such as finance and real estate.

When I graduated from a London university, I knew that Bloomberg provided the most-used data for investment banks and was in the forefront of delivering the best data.

I was then rather shocked when I started to work for Bloomberg in London and got my hands on a Bloomberg terminal and its interface. It reminded me of Microsoft DOS from the 1980s. This was a multi-billion-dollar company that was heavily investing in software development, and it was nowhere near to Electronic Arts' software, or other software from start-ups. By working mainly with fixed-income clients lending to real estate investors, I have seen that the property industry can be entered easily by utilizing technology.

This is what I did, by starting my own real estate firm in Germany where I have used search engine optimization strategies to become top-ranked, generating new leads globally and selling properties using technologies only. My ability to use six languages probably helped. We produced real estate and virtual tour videos, advertised via social media, and sold properties without ever seeing or talking by phone with some of our clients—all by using technology and the newest software. We have outperformed market players with 50 years or more in the market.

Due to our innovative approach, we have been awarded with prizes from one of the top universities in Germany. We also had a tech lab in a start-up hub, which grew my network of programmers and full-stack developers. This has also accelerated the implementation of technology in our daily processes. At one meeting with a real estate client in our office, we were asked for a pen, but we had forgotten where we had stored them. We were using only iPads and DocuSign for contracts.

After five years, I received an offer from the largest housing association in Europe to use technology to optimize a large real estate portfolio. When I began in London, the residential portfolio was worth around EUR 60m, which I have optimized to EUR 70m and sold for around EUR 95m.

After that transaction, I worked for Europe's largest and leading real estate advisory for construction and building engineering, where I was the head of real estate investment in London. I now have worked on transactions worth about $3.5 billion all over the world.

By collaborating closely with the in-house research and development department, we were able to consult on and develop the world's smartest building, "The Cube," in Berlin and develop the Amazon Robotics logistics center in Europe.

Apart from my ventures, I am heading the investment department of a very rapidly growing investment and asset management firm, where I am using my PropTech knowledge to increase the number of transactions, optimize processes, improve communication, and make our assets smart.

PropTech Solutions

PropTech is the abbreviation for property and technology and has been widely accepted as the synonym for the digitized and automation-focused transformation within the real estate industry.

So, even if it might never be possible practically, the ultimate aim should always be to try to replace all intermediaries with technology or digital solutions and make all software solutions compatible.

If you believe that this sounds a little bit like a utopian vision, just think about firms like HomeAway or Booking.com. A host in one part of the world advertises a property on a digital platform by answering standardized questions. The traveler or guest from a different part of the world searches for a property on the other part of the world by answering a similar questionnaire. An algorithm programmed by the developers of the platform matches the ideal fit. This platform then replaces the host's broker and the guest's broker and the solicitors, as the deal follows the firm's terms and conditions. This strategy is possible because people are willing to use technology and travel globally.

If you now say this is possible for structures that can be standardized and pre-programmed, but is more difficult in a world where every property and real estate transaction is different, I fully agree. Every new situation is different and must be re-thought and tackled in a unique way. Therefore, human intervention is needed, as we can think through a problem to find a solution and optimize that solution.

However, we have reached the time where we are able to envision self-driving cars, and Google already is using artificial intelligence. Self-driving cars need milliseconds to navigate new and unknown situations and can recalculate a solution in every millisecond. Now that we have invented this kind of technology, we should also look for ways to use it in our real estate industry, even though it is making so much money that some think we do not need to experiment.

Let's face it, and be more honest with ourselves: at the moment, it seems like everything is going well and we do not need to make any changes. Hence, we will not make any changes, as it can be difficult to adapt to a new environment.

But don't be fooled. If you, as a real estate expert in your field, choose to not make any changes, recognize that someone else surely will make changes by trying to find a better solution to the same problems you face, and you might lose your job or competitive edge in the not-too-distant future.

This book is designed to help you face the current and upcoming challenges in the real estate industry globally, and provide hands-on and applicable solutions in the form of brief information about start-ups that are solving day-to-day problems and issues. The PropTech industry is in its infancy, and we can expect several problems as it grows and matures. In this book, I am questioning our need for all the intermediaries we use today, and posing this question: Can it be done faster and more cheaply? In so doing, I will be focusing on the same three attributes throughout all stages of a property:

1. Transparency
2. Speed
3. Cost

By implementing newly available technological solutions that are providing greater transparency, and better, faster, and cheaper solutions, it will be possible for current real estate market players to increase profitability. This will allow them to be ready to fight the upcoming battles with new technology-focused players now entering the real-estate world who want to provide better solutions.

For commercial real estate market players, this book will give a good overview of a number of firms that are solving problems that those market players must deal with every day, but may not have the time to research solutions for.

Residential landlords and tenants will learn how professionals have dealt with daily property issues, and how they might use new solutions from PropTech firms, which cost often a fraction of what commercial real estate market players are paying. Landlords can learn about new ways to manage or improve their properties, and tenants can discover the benefits of applicable technological solutions that they can suggest to their landlords.

PropTech participants find their opportunities by understanding the pain real estate participants encounter by using outdated methods to solve problems. These new digital- and technology-focused firms can find better solutions or customize their digital products and services to the property industry.

As this book covers all stages of the property lifecycle, I recommend you read through the book and then rethink the problems you are facing in, on, and around your property. Always refer back to the section that is most relevant. After you have read this book once, you will remember that there was a way to tackle the problem either faster, cheaper, or more efficiently.

Transparency

Transparency during the construction phase

Do you know what has been actually constructed, or which materials have been used? I believe that even architects and engineers sometimes do not know for sure all the components that have been used during a development. For example, where are the electrical cables that have been placed between walls, and are not visible? How would you fix an electrical problem in a hundred-year-old building? These people are very educated in their field, and there is a very good chance that they can assume where a problem might be occurring and how to access and fix it. But at best, this would be an educated guess, and potentially an expensive and time-consuming one at that.

Such uncertainty will largely go away in the future given the introduction of Building Information Modelling (BIM), which I hope and expect will provide greater transparency in the construction process in the future. Building Information Modelling is a process where a building to be constructed first is created and represented digitally. The data within this digital model can be managed, extracted, exchanged, and shared among many different stakeholders before the construction process starts. Even after the completion of the construction project, the digital model will be continuously updated so that the newest owner of the building has the latest information of what he has bought.

Transparency during a transaction process

Today, the typical transaction process features several stakeholders who communicate by phone, messaging, and email. In a non-commercial real estate transaction, there is no central place where documents are saved in a digital form. Lawyers and solicitors are mainly using Word and PDF formats. Banks have their own software, or sometimes also use Excel. If you are lucky enough to have very good real estate agent, he might be the one who is chasing down all the different stakeholders and informing you about the status of the transaction. If you're not that lucky, then you must do all the chasing. This entire process demands a lot of time and energy, and also costs a lot money.

So, you might wonder, why is there not another solution? Why are people still doing it in the old way?

Well, the answer is very simple. It has evolved in that way over time and is still making money for all participants. Probably the most important factor is that every stakeholder in this process is earning enough money to continue doing it in the old-fashioned way. There seems to be little motivation to streamline the process.

Transparency during the asset management phase

During the asset management period, the owner of the asset is holding the property and ensures proper maintenance. But there is a discrepancy between what the tenant knows and what the landlord knows, and there is a conflict of interest. The tenant is the person who knows the property best during the time of occupancy, and he is also the one who faces all the problems first. The landlord is the one who sees the problems last, but is the one responsible for executing and paying for remediation most of the time.

So, the way the repairs have been conducted has been simple: The tenants see a problem, defect, or malfunction needing to be repaired and report it to either the facility manager, the property manager, or the caretaker of the property. The caretaker either fixes the problem himself or, if the cost of remedy is higher than a certain threshold, reports it directly or indirectly to the landlord. The landlord then decides whether to fix it or to leave it.

This entire process from the discovery of the problem until it reaches the landlord could have taken weeks or months. Then there often is further delay in making a decision and getting the problems fixed. Lots of time and money can be lost during such a process. The ideal situation would have allowed the tenant to not wait until the problem occurred, perhaps by having the ability to alert the landlord in advance that there might be a problem which could be fixed before a breakdown resulted in a costlier repair.

Today we can use technology to obtain data to analyze building and systems performance and indicate when maintenance is needed or sound an alert for a potential risk or breakdown in the short-, mid-, or long-term. The landlord does not need to rely on past data indicating a rough life expectancy of a building or its components. Instead, real-time data from sensors within the building, combined with sophisticated algorithms, can predict a breakdown (e.g. of a lift in a building). So, before this problem happens, the landlord can address the problem and can reduce the cost of repairs and maintenance.

Speed

Speed can be defined as the rate at which someone or something moves, operates, or is able to move or operate. In the real estate world, the word "speed" probably didn't exist. The entire real estate ecosystem moves so slow.

Velocity of raising money and buying real estate

The passage of time in real estate starts with raising money, which requires lots of meetings and appointments and extends into the construction phase, where it takes ages until a building has been completed.

It then takes more time to transfer ownership from one party to another during a transaction process. In a lot of countries, it is still believed that buying a house is an investment for your life. Hence, a mortgage is issued for 20 to 30 years. The word "mortgage" derives from the word mortem, which stands for "until you die." This is not the case anymore.

The speed of constructing a building

The Empire State building was built between 1930 and 1931 and was the largest building at the time. The Burj Khalifa was built between 2004 and 2009 and is the largest building today. So, it took only a bit more than a year to build the largest building in the 1930s, and today it takes more than five years to build the largest building.

Of course, the Burj Khalifa project was so complex that it required a lot of coordination by the individual stakeholders and a lot of planning. One of the biggest problems during the construction phase was the errors and mistakes, which always results in a delay. Luckily, we are in a time when we can utilize Building Information Modelling, virtual reality, augmented reality, and live cameras to better plan and coordinate the construction.

Cost

A property transaction is expensive

A change of ownership of a property is one of the costliest events during its lifecycle. Every stakeholder involved in the transaction process must be paid. This starts with the lender who is often charging a fee to provide the funds, and ends with a property manager who is charging a setup fee to add the property into his portfolio.

So, a list of participants during the transaction can be as follows: the bank, the surveyor, the buyer's broker and lawyer, the seller's broker and lawyer, the notary, and the city council that is registering the transfer.

Depending on the property location, type, and size, the number of participants can vary. Often the fees are pegged to the price of the property, where all participants are charging a percentage instead of a fixed fee. In Germany, for example, a private buyer of a residential property who involves a broker can roughly estimate to pay up to 15% in addition to the selling price.

With so many different participants during the transaction process, it is obvious why this transaction process is so costly. This might be another reason why property is held for such a long time before it changes hands again.

A number of start-ups are working on technological solutions to improve this situation. Their aim is to reduce the time for communication and try to optimize transparency. There are digital platforms where it is possible to designate roles within the transaction so that participants can see in real time the progress being made toward accomplishing the required work.

Other start-ups are focusing on replacing the intermediary that is the real estate broker. They are using algorithms to match the properties with investors or buyers. This could lead to omitting much of the marketing process, where the broker is pitching and convincing the vendor to sell and the investor to buy. Instead, they are using pictures, videos, virtual tours, and augmented reality to make a property so attractive that the buyer does not need to be persuaded further. By being able to scale this technology, it is possible to charge a lot of people a small fee instead of charging large fees from a few transactions.

Cost of construction

In the past, determining the cost of construction was a matter of experience and estimation. This is not satisfying for an investor with a finance background who would prefer an exact figure for the cost to construct a building. The problem is not that the engineer, architect, or the contractor could not provide an exact figure. Instead, during the construction, any number of variables can change, altering the validity of the initial cost projection. Sometimes even the investor comes up with new changes during the construction phase, unaware of the cost implication.

The construction industry has taken on two new ways to improve the construction process. The first way is copying the automotive industry, which can calculate the cost of building a car down to the penny, mainly because the entire production is happening within a factory, which is lean and automated.

This has given rise to the trend in the building industry of constructing parts of the building in a factory, then shifting it to the construction site for installation and final touches. Some companies go one step further and construct the entire building in the factory and transport it to the land to be placed and used without any further requirement of construction.

Computer-Assisted Design allows for a lot of changes and amendments to be visualized before ground is broken, as does factory pre-assembly. But the problem lies in the compatibility and translation of the data into the required information necessary for non-architects and others. Due to the evolution of technology, it is now possible to fully plan and visualize a building before starting the construction. This is the second way. It is possible to show all participants final results of the development project from every angle in a three-to-five-dimensional virtual model. This means planners can show the process and progress of development, the components being used and installed, and finally the implication of changes in terms of cost and time. In this way, all contractors can visualize the installation of their components within the building and also understand problems or errors virtually before they happen in reality. Furthermore, the architect and the civil engineer can show the investor, who might be from the financial industry, the implication of cost and time for any changes he suggests before, during, and after the construction period.

This doesn't mean that every construction project can be calculated with hundred-percent certainty, there can be events that delay construction, which involves a very long supply chain and inconsistent weather, with one glitch compounding into another. It can always happen that someone discovers a not-anticipated World War II bomb on the construction site, where the construction has to be stopped immediately. Or there might be accidents which can cause delays in the completion of the development project.

2

Match-Making

"The future depends on what we do in the present."

–Mahatma Gandhi

Marketplaces

The place where there seems to be the most innovation, which is easy to see and to understand, is in the broad category of the age-old art of bringing together buyers and sellers, and leasers and renters. So, we start with this category.

This has long been the role of the broker, mainly the "local" broker who made his money with connections; knowing properties, neighborhoods, buildings, and the available land; and money spent on the marketing side to draw in people interested in buying or renting.

That has evaporated rapidly and is being almost wholly replaced by digital marketplaces. This chapter breaks this down into various sub-categories:

- home sales
- residential rentals
- pop-up spaces in malls
- commercial properties including office buildings
- all types of raw spaces from warehouses to storage facilities.

The goal is to outline all companies who are offering new "marketplaces" designed to bring together buyers/renters with sellers/leasers. We focus in this chapter on the function within the existing structure that is under attack by technology: the brokering function.

Innovative Firms

Real Estate Agencies

Company Name: Triplemint
Website: https://www.triplemint.com/
Target user: Renters, buyers, and sellers
What they do: This company from the United States is a digitized real estate brokerage company, which uses massively big data to match search criteria and properties. With a large and growing database of potential clients interested in leasing, buying, or selling, predictive marketing campaigns can be very effective. A strong advantage of this platform is that it is offering the clients off-market transaction opportunities.

Company Name: Yanport
Website: https://www.yanport.com
Target user: Property sellers
What they do: Listing properties with every agent and on every platform can sometimes make a property unattractive. A potential buyer might think there is something wrong with a property that is advertised everywhere. On the other hand, healthy competition can increase the efficiency of the marketing and sales strategy for a property. This company from France uses a strategy where it gives permission to advertise the property to only three property brokers. The property brokers can monitor the competition digitally to heat up the marketing and sales efforts.

Company Name: Purplebricks
Website: https://www.purplebricks.co.uk/
Company Name: Yopa
Website: https://www.yopa.co.uk/
Company Name: Reali
Website: https://www.reali.com/
Target user: Buyers and sellers
What they do: A property broker may use a number of marketing efforts, which can be very inefficient or ineffective. A lot of information about a property is known best by the property owner. These firms have created a hybrid between a standard physical property agency and the digital property brokerage. The online marketing and advertisement is being taken over by the broker, whereas other activities such as opening the doors for the viewing can still be conducted by the seller. Making or receiving an offer can still be done online. In the UK, this model has shown to be very successful.

Company Name: Redfin
Website: https://redfin.com/
Target user: Buyers and sellers
What they do: This company is a real estate brokerage, but it has digitized all the processes. Furthermore, it focuses on delivering services, instead of closing the sale. Employees are not being paid by commission, instead by a fixed salary. Customers can use virtual tours through the website, which reduces the task of showing the property to a client.

Innovative residential property buyers

Company Name: Opendoor
Website: https://www.opendoor.com/
Target user: Owners and sellers
What they do: Reducing the stress of selling a home is the big advancement in the real estate industry. This company from the United States is using its digital facilities to make the selling process as simple as possible. It buys the property from the seller at fair market value and enables all work to be conducted at the home of the seller. The seller has the certainty that he is not selling under market value and is not exposed to the transaction stress. The inspection and the signing with a notary can be conducted at the property.

Company Name: Nested
Website: https://nested.com/
Target user: Residential sellers
What they do: If someone is already the private owner of a property, it is even more stressful if they decide to buy a new property and to sell the current property simultaneously. This company from the United Kingdom uses a digital platform to offer landlords certainty when moving on to the next property. After evaluating the property for sale, it enables the landlord to withdraw an advanced payment if the landlord requires it to pay towards the new property. This advanced payment can even be taken out before the completion of the sale of the current property. It makes the landlord virtually chain free.

Company Name: Knock

Website: https://www.knock.com/

Target user: Residential sellers

What they do: This company understood that it's easier to complete refurbishment work within a property once it is empty. It allows its client to move to a new house before the sale has been completed. Once the new property has been found, the refurbishment work in the new property can be executed. After that, the landlord can use the new premises. Finally, the old property will be refurbished and sold at maximum price. The company completes all necessary work. This means the firm offers a guaranteed safe price. The seller moves out, the refurbishment work is executed, and finally the property is offered to the market and sold.

This strategy can work well due to the use of data to determine a price the company can afford.

Market places

Company Name: RealX.pro
Website: https://realx.pro/
Target user: Commercial investors
What they do: A number of commercial real estate transactions now are happening behind closed doors. These off-market deals are very attractive to real estate investors. Previously, real estate brokers had a database of potential clients interested in buying. Once a new property portfolio was for sale, the broker would attempt to match potential buyers and offer the deal in private. RealX.pro is offering a digital platform to bring those brokers, investors, and the financing bank into one place. The transaction material can be created through this platform by the brokers. As the material is standardized, investors and banks can evaluate the potential deal much faster and more efficiently.

Company Name: 21st Real Estate
Website: https://www.21re.de/en/
Target user: Commercial investors
What they do: This company from Germany has digitally transformed the entire transaction process for commercial real estate and real estate portfolios. The firm uses an algorithm and big data, and has created a growing database of properties in all of Germany. Registered brokers can place their properties for sale into this digital platform to attract the most suitable buyer.
Real estate investors can define the exact search criteria for the commercial property or the property portfolio on the digital platform, which uses its algorithm to find the best match and gives an overview of potential deals which might be suitable for the investor without a long search. Using this platform, the investor avoids the lengthy process of going through an analysis of hundreds of properties. Furthermore, it enables the investor to compare deals and see a score, which tells whether a property is

over-or under-valued. Finally, it enables the investor to complete the transaction with digital support and the use of workflow management, where the roles of individual stakeholders during the transaction process can be defined and integrated into the system.

Company Name: Envelope
Website: https://envelope.city/
Company Name: LandInsight
Website: https://landinsight.io/
Target user: Architects, investors, surveyors, and more
What they do: It has been always a hidden secret to find suitable land that can be developed into a profitable building. These companies provide a software solution that enables developers and other real estate participants to find land, obtain data and information about the zoning, analyze it, visualize it, and run scenarios for potential development. It is also possible to receive market alerts to react quickly in order to find new potential deals.

Company Name: ShareDnC
Website: http://www.sharednc.com/
Target user: Office owners and renters
What they do: This company focusing on Germany and Austria provides a marketplace for flexible offices, shared offices, and co-working spaces. Companies can maximize their office space by renting out unused desks. Landlords can rent out office space on a short-term basis for a temporary purpose. Additionally, business centers and other co-working spaces are also available on this digital marketplace. The aim of the company is to provide a database of offices for any budget, any city, and any size.

Company Name: Stowga
Website: https://www.stowga.com/
Target user: Warehouse owners and renters
What they do: Due to the success of online retailers, warehouses and other logistics properties have emerged as a demanded asset class. This company provides a marketplace for warehouse spaces through a global network of providers who can offer space below market price, and supports the completion of the administration process.

Company Name: Storage Share
Website: http://www.storage-share.nl/
Target user: Warehouse owners and renters
What they do: This company from the Netherlands is providing a marketplace where owners of unused storage space can market their premises, and users in need of storage facilities are able to use it at a lower price. Therefore, real estate owners can maximize utilization.

Company Name: The eLocations Ltd.
Website: http://www.elocations.com/
Target user: Retail space owners, property managers, and renters
What they do: This company from Austria is currently accumulating retail store data globally. The aim is to create a marketplace to connect brands and retail companies with landlords and property managers who are letting out their premises. Such a global marketplace can be very beneficial for companies thinking about expanding to a different geographical location without the time or means to continuously travel to conduct viewings. It painlessly enables its landlords to promote their premises globally.

Company Name: StoreFront
Website: https://www.thestorefront.com/
Target user: Retail space owners, property managers, and renters
What they do: Short-term lets are not only popular within the real estate residential and office sector, but also in the retail sector as it becomes more and more popular to rent short-term. This company provides a digital marketplace where owners of retail store space can find new tenants on a short-term basis. Major brands, designers, or even artists can find retail space where they can market their products in a premise on a busy street, within a boutique store, or even at a street fair. The digitization makes this marketplace scalable and accessible to internationals, who can rent a retail space on the other side of the world.

Company Name: Breather
Website: https://breather.com/
Company Name: Spacebase
Website: https://www.spacebase.com/
Target user: Artists, renters, property managers, office and retail space owners
What they do: Companies, artists, and the creative industry might require an office or work space irregularly to exhibit their work or to conduct a business meeting in a different location. These companies provide a digital platform to book an office space, venue location, and other working areas to rent out on a short-term basis without any commitment.

Company Name: Popertee
Website: http://popertee.com/
Target user: Retail space owners, property managers, and renters
What they do: Major brands such as Coca-Cola or Virgin are conducting marketing campaigns on a regular basis that require properties on a short-term basis. This company provides a platform where brands can identify suitable premises, which can

be an entire store or just a pop-up within another store, in which to conduct the marketing campaign. The platform enables them to reserve and book the property and analyze the impact of the campaign.

Company Name: Go-PopUp
Website: http://www.gopopup.com/
Target user: Retail space owners, property managers, and renters
What they do: This company provides a retail market platform to offer and find premises all over the world. It is possible to get access to locations within a shopping center, a pop-up container, or any other interesting location to conduct a marketing campaign or exhibit products. The platform's emphasis is to make booking and offering retail space as easy as booking a hotel room online.

Company Name: Commuty
Website: https://www.commuty.net/
Target user: Car park asset managers and property managers
What they do: This company from Belgium understood the problem of commuters who are travelling downtown into dense city areas. It connects car parks with travelers and commuters, enabling them to book a parking spot in advance and avoid the search for a parking space during rush hour.

Furthermore, it provides a smart parking management system, a car pooling system, and recommends usage of bicycles by connecting cyclists for bike sharing. Owners of car parks can advertise their parking spaces, which may have been empty in the past but now can be fully occupied.

3

Finding the Money

"If you would know the value
of money, go and try to
borrow some."

–Benjamin Franklin

In this chapter we will analyze the various platforms that are growing and getting more sophisticated to finance the real estate business, starting with firms that are assembling big data to provide buyers and investors with a better grasp of market value, trends, and suitability, to platforms allowing small investors to buy a piece of an aggregate project, to platforms linking buyers to lenders.

This all can further "flatten the Earth" by allowing investors to reach beyond old geographical boundaries to understand distant or foreign markets and trends and invest confidently, while also allowing companies of all sizes and types to better home in on locations that best serve their interests.

On the residential side, home buyers now can effectively shop from their home offices for mortgages, properties, and all that goes with it with new understanding, wider reach, and more confidence about everything from a home, to a vacation rental, to a storage unit.

Raising capital, the traditional way

In order to buy a flat, house, or even an entire building, you either need to have all the required cash available, or be able to raise the funds. The standard route is that you save up some money for a down payment and apply for a loan at your local bank, as you usually don't have enough cash available to buy the property outright. Hence, you are raising capital—either through equity or via debt.

The traditional way to obtain a mortgage is a very lengthy process. First, you need to save up enough money to pay the deposit. This is already a big obstacle, which most people in their twenties in major cities of the world are not accomplishing anymore, as the prices of properties have appreciated enormously. Afterwards, you should compare the prices and interest rates of the mortgages you are applying for. Some people just go to their local bank and get a mortgage, which is not ideal, as they are skipping the entire process of comparing mortgage rates and conditions. Then there is the entire application process, which might start online, but often ends offline. An example could be as follows:

1. Make an appointment with your local bank.
2. Go to your bank with all the documents they have requested as a print version.
3. Fill out several forms and questionnaires and sign the documents.
4. Wait until you receive a response from the bank via post. Check all the documents again and sign the remaining application forms and send them back to the bank. Finally, wait for a response from the bank, which you will receive by post.

This process might take a few days, if you're lucky. However, this can also be a very lengthy process, where you must wait a few months until you get a decision. The entire process is not very transparent, and it is not fully digitized for the end-customer, especially the commercial property loan borrower.

Capital raising the digital way

The solution for the problems just mentioned is bridging the gap of information between what the customer needs and what the lender is providing.

The speed, efficiency, and effectiveness of providing documents, data, and information from the lender to the customer, and from the customer to the lender, can be improved with technology. Finally, it is worth considering new solutions provided by start-ups that are omitting an intermediary with the use of technology.

Looking at the capital-raising methodologies, we are differentiating between equity capital and debt capital. We are also distinguishing between the digital industry of raising debt as the lending tech market and the digital industry of equity raising as equity crowdfunding.

Lending tech

Within the lending tech industry, we can differentiate between data companies that are collecting and providing aggregated data to facilitate the mortgage application, and firms that are online lenders and underwriters that have digitized the mortgage application and lending process.

These data analytics companies provide crucial input for commercial lending, but also for non-commercial needs. The creation and the collection of data has increased exponentially in recent years. These companies are collecting this information, analyzing it, and providing aggregated results that are the best matching solution for businesses and for the end user.

This means that no individuals or large investment management companies need to research newspapers and magazines or make phone calls to brokers to obtain the data they need to invest and deploy their money. Artificial intelligence (AI) and algorithms are enhancing this method of analyzing the data—quickly, efficiently, effectively, and without errors. For example, as the ability of the software to read scanned documents advances, the process becomes smoother and quicker.

The digital loan or mortgage lenders that are providing capital through an online application process are using information from the data firms to create new concepts to speed up the application process. For example, a small construction company can borrow at a much lower interest rate than they would have received if they limited their outreach to a local bank. The construction company provides information about the financials of the project. The software searches hundreds or thousands of lenders digitally to find the most suitable loan provider. The efficiency of the process allows the construction company to focus on the core business, which is developing a building, instead of going through a lengthy search and application process.

The interesting part about this is that the construction company is not only getting money, but has options beyond a bank or a financial institution. Very often, they can get money directly from the investor. This means the middleman has been undermined and the cost has been reduced.

Equity crowdfunding

In the past, investing in properties meant you were buying your own home, a rental property, or buying shares in a real estate investment trust or open- or closed-ended fund. Investing in your own home or a buy-to-let property meant a lengthy process of investment and required large sums of money. Investing in a commercial real estate fund meant that you needed to pay high fees in a non-transparent management structure that was taking care of the real estate. This was absolutely the same for the borrower, who could only rely on banks or large institutions that aggregated the sums from individual investors and forwarded this with a high fee to the borrower.

A real estate crowdfunding platform collects money from a number of different smaller investors to assemble the minimum investment amount criteria. Hence, the platform is able to access high-yielding investment options that were previously accessible only to accredited investors. For private and non-accredited investors looking to diversify their real estate investment portfolio, digital and online real estate crowdfunding platforms offer a fast, easy, frictionless option that serves as a 21st century bridge connecting them to previously inaccessible property investment opportunities.

Furthermore, for both real estate investment sponsors and borrowers looking to raise joint venture equity with flexible capital, these platforms provide access to equity and debt that would likely take a long period to secure through the more traditional, slow, and restrictive means I have described. These traditional and conventional ways of investing in a share of a real estate fund were Real Estate Investment Trusts (REITs), and Open- and Closed-ended funds, which had a number of intermediaries who all received a commission the investor had to pay. Hence, the actual profit from the property was reduced.

The aim of the real estate crowdfunding platforms is to replace those intermediaries by using technology, through which the investor can see how, where, and with whom his money is working.

Hence, whether a novice buyer, an accredited or institutional investor wanting to diversify his portfolio across real estate asset classes, a borrower in search of a flexible loan, or a real estate investment sponsor looking to raise joint venture equity, these platforms can bring real estate investing into the 21st century. As with the introduction of crowd investment platforms, we might have a solution for transparency, efficiency, effectiveness, cost, and speed. This means the lender can invest directly in a project via an online and digital platform because there will be only one intermediary, which is the digital crowdfunding platform. Digital means there is a standardized process and uniform requirements to borrow and to invest online. The more efficient digital platform in this case replaces, for example, the bank.

Selling shares of existing properties

In the past, a property owner was able to refinance assets and receive money for the loan. In this case, the owner accepted further debt on this property, which he had to repay. New companies are providing an option in which the owner does not need to take on debt, instead the company buys shares of the asset. These companies are expecting lower operational and capital expenditures because the owner, who is living in the property, will maintain it better than tenants with no equity invested in the property. The owners of the properties might lose on price appreciation for the stake which they have given up. On the other hand, the investing companies have downside risks when the prices are depreciating.

Mezzanine funding for developments

Mezzanine capital is required during a transition period, which is rather short. This means during the construction, acquisition, or sales period. It is also riskier because a number of unexpected circumstances can occur that can change the outcome of the project. However, once the project is completed, the returns are higher. In the past, this capital was only raised through banks or financial institutions that specialized in such situations. Today, the digital platforms that are providing capital to those projects also have a marketing effect. Investors can see the project, follow the development, and either buy or recommend the acquisition of the property when completed.

Innovative Firms

Equity crowdfunding

Company Name: ICapital Network
Website: http://www.icapitalnetwork.com/
Target user: Institutional investors, hedge funds, private equity companies, and institutional and accredited investors
What they do: This company provides a digital financial technology platform and aims to deliver access to real estate and alternative investments to institutional investors, hedge funds, private equity companies and accredited high-net-worth investors. The platform provides access to capital and enables qualified investors to start to invest from a minimum of $100,000. The usual minimums would otherwise be several multi-million dollar investments. The technology enables the investors to see the investment pitch of the fund managers through an online video roadshow. Furthermore, they can assess offering documents, information, performance, and track records in order to finally invest directly online through their platform. The firm provides a turnkey solution for alternative asset managers to raise millions of dollars from investors via this platform. As this platform is online, it is scalable to a global reach for the investor and the borrower.

Company Name: Source Central
Website: https://www.sourcecentral.co/
Target user: Institutional investors, asset managers, and limited partners
What they do: The company is a cloud-based platform providing information, documents, data, and reports for institutional investors, asset managers, and limited partners. Furthermore, it brings together institutional investors with asset managers and the real asset investment opportunities.

Company Name: BrickVest
Website: https://brickvest.com/
Target user: Institutional and accredited investors
What they do: This firm provides a commercial property investment platform that links investment opportunities directly with investors. Its focus is the commercial real estate market, which previously was only accessible to large institutions and investors. Through its digital approach, it is able to provide a transparent alternative to Open/Closed-End Funds, but without high fees and mishandled clashes of interest between stakeholders.

Company Name: Realty Mogul
Website: htttps://www.realtymogul.com/
Target user: Non-accredited, accredited, and institutional investors
What they do: This company aims to simplify real estate investing through the use of technology and is a digital peer-to-peer lender for real estate investments. It brings together property investors and real estate companies with a need to raise debt or equity capital. Through the platform, it is possible for non-accredited, accredited, and institutional investors to invest in real estate deals. Those investors can review documents, data, and information, and invest online by digitally signing legal documents. Borrowers can raise debt and/or equity funds by going to an online application.

Company Name: Cadre
Website: https://cadre.com/
Target user: Accredited and institutional investors
What they do: This company provides a technology-enabled real estate investment platform and is emphasizing an efficient due diligence process with the use of digital data and a fast application process. Sophisticated investors can start to invest with US$100,000. The management board consists of former investment bankers whose aim is to increase efficiency during the transaction process. Goldman Sachs has committed $250 million in money collected from private-wealth clients.

Company Name: CoAssets
Website: https://www.coassets.com/
Target user: Non-accredited, accredited, and institutional investors
What they do: This company is listed on the Australian Securities Exchange (ASX) and provides a crowdfunding platform in Southeast Asia. Property investors can connect with developers and small- and medium-sized enterprises (SMEs) from Singapore, Malaysia, Indonesia, Australia, and China.

Company Name: Prodigy Network
Website: https://www.prodigynetwork.com/
Target user: Non-accredited, accredited, and institutional investors
What they do: This company's crowd-investing model is providing access to commercial real estate investment opportunities. Their offices are located in New York City; Miami, Florida; Bogota, Colombia; and Montevideo, Uruguay and they are reaching out to 6,200 investors, major banks, and operators globally. Their global projects have reached a value of more than $850 million and they are well placed (e.g. in Manhattan).

Company Name: RealtyShares
Website: https://www.realtyshares.com/
Target user: Accredited, institutional investors and borrowers
What they do: This company provides a property crowdfunding platform to raise capital to finance debt or equity needs. It uses technology and its online platform to bring investors and borrowers together, therefore it omits the middleman, bank, or broker. It raises funds for residential and commercial real estate projects and also has a strong focus on gap financing, senior debt, and second lien loans as a subordinated financing solution to increase leverage.

Innovative funding providers

Company Name: Walliance
Website: https://www.walliance.eu/
Target user: Non-accredited, accredited, and institutional investors
What they do: This company provides the first digital Italian real estate investment platform in the form of crowdfunding. The company is part of a holding in the third generation of the family business, which now made the move from offline to digital investments. With €500, investors can buy shares in a property. The portal is authorized by the Italian companies and exchange commission and will do a thorough due diligence process before accepting any investment projects from borrowers. A large portion of investors are coming from Italy, interested in investing in their own country and also abroad.

Company Name: Exporo
Website: https://exporo.de/
Target user: Non-accredited, accredited, and institutional investors
What they do: This German-based start-up provides a digital crowdfunding platform providing investors the chance to participate in well-analyzed real estate projects. The firm combines asset-backed investments and real estate project experience. They have a strong focus on funding development and construction projects and will provide bridge and mezzanine capital.

Company Name: Shojin
Website: https://www.shojin.co.uk/
Target user: Non-accredited, accredited, and institutional investors
What they do: This company, which is based in the UK, has been operating since 2009 and made its transition to a digital investment platform. It provides a UK-focused real estate development and investment solution, enabling equity funding for property developers, and co-invests with investors in every single project. The firm arranges buy-to-let, bridge and mezzanine loans, and the core product focuses on funding for construction and development. The firm has expertise in construction, finance, and project management, which creates attractive investment structures.

Company Name: Housers
Website: https://www.housers.com/
Target user: Non-accredited investors
What they do: With only €50 per month, it is possible for the average investor to deploy funds in real estate. The firm provides a digital platform where investors can invest in traditional buy-to-let properties or development projects.

Company Name: Property Partner
Website: https://www.propertypartner.co/
Target user: Non-accredited investors
What they do: This firm was founded in 2014 and is based in London. It is a real estate crowdfunding platform that enables anybody to finance residential real estate digitally. It also provides an exchange, so investors are able to trade their shares. The platform also shows rental income and capital growth. Its mission is to democratize real estate investments. The fully digitized real estate platform enables people to invest in commercial real estate, student accommodation, and residential properties. It has a very strong focus on buy-to-let properties. One of its strengths is that it has a strong team with tech and property backgrounds and a reputable board of advisers.

Company Name: CapitalRise
Website: https://www.capitalrise.com/
Target user: Accredited and institutional investors
What they do: This company provides a digital online platform for high-value properties. Its competitive advantage lies in the speed of the transaction, as it aims to provide funds to the borrower within 24 hours and can complete the funding process in 20 days or less. Investors are able to earn 8% to 12% per year, and the capital raisers can obtain funds for senior, stretch senior, mezzanine, preferred equity, bridging, development, and sales period loans.

Company Name: Brickowner
Website: https://brickowner.com/
Target user: Non-accredited investors
What they do: This company is based in the UK and has a team with a very strong background in real estate. It provides a real estate crowdfunding platform to enable investors to invest in UK properties with a strong growth potential.

Company Name: Property Moose
Website: https://propertymoose.co.uk/
Target user: Non-accredited investors
What they do: This company based in London provides a fully integrated digital property crowdfunding platform. Investors can invest in a diverse range of investment opportunities with minimal investment capital of £100. It aims to democratize investing by giving a non-accredited investor the opportunity to invest in high-value investments. The digital investment platform enables the investor to buy in expertly sourced buy-to-let opportunities, lend directly to or alongside carefully vetted property developers, and sell shares on one of the world's first property trading markets. The property investment platform provides either a capital yield, price appreciation, or even both, and the investor can decide to stay with an investment or sell the shares. After the end of the fixed investment period, the investor can vote to sell the investment on the open market, or retain shares for a further fixed term. This process is digitized, and the buyer receives voting opportunities via email, which means the investor can vote whether or not to sell the building. By selling the building, the investor will receive a corresponding share of the sales price and, by keeping it for another period, the investor will continue to receive a share of the monthly rental income.

Equity sharing providers

Company Name: Point
Website: https://point.com/
Target user: Property owners
What they do: This company provides a platform enabling residential property owners to sell equity directly to investors. Share prices correspond to the value of the property and can fluctuate.

Company Name: Unmortgage
Website: https://capital.unmortgage.com/
Target user: Residential property buyer
What they do: This company aims to bridge the gap between renting and owning. It enables the buyer of the home to buy a part he can afford and rent the rest until he can afford to buy it at a later stage. It is similar to a shared ownership scheme, but without the requirement of taking out a mortgage. The buyer receives funds to purchase his home worth up to ten times their income with only a 5% deposit. Through the digital platform, the investors have access to a selection of mid-market family homes in reliable areas producing an inflation-linked net yield of 5%. The tenants are directly invested in their own home and have an interest in maintaining their property and keeping costs low.

Company Name: StrideUp
Website: https://www.strideup.co/
Target user: Residential property buyer
What they do: This company provides a shared ownership scheme, in which the buyer can buy a property and own the part of the property's equity that he can afford. He pays rent for the equity he does not own. The aim is that he saves up money to buy the rest of the shares of the property.

Innovative data collection firms

We have provided a list of firms who are providing data analytics by collecting and evaluating information before, during, and after the loan application process to forecast creditworthiness reliability and behavior of the applicant. The aim is to finally provide objective solutions for the lender and the borrower.

Company Name: Credit Sesame
Website: https://www.creditsesame.com/
Company Name: Equifax
Website: https://www.equifax.co.uk/
Target user: Mortgagers and borrowers
What they do: These companies are credit agencies and loan management platforms that collect data from customers to analyze and provide the best loan according to their creditworthiness and credit rating. So, the customer will be offered products to be able to apply for a mortgage or a loan that has a high potential to be accepted. Additionally, the customer will be able to monitor his credit rating on an ongoing basis. As data has been accumulated over a lengthy period before the application process, the predictions and scenarios can be very accurate. Furthermore, this knowledge can be provided to all lenders and is not restricted to one bank with whom the client had a long relationship.

Company Name: Yodlee
Website: https://www.yodlee.com/
Target user: Banks and financial institutions
What they do: This company collects, aggregates, and analyzes data to predict the next move of the client's end customer. It works with banks and lending institutions to check the creditworthiness of the customer to decide whether to lend or not, and finally to provide the best product for the customer.

Company Name: Canopy
Website: https://findyourcanopy.com/
Target user: Private landlords and tenants
What they do: The most important business client of the landlord is the tenant paying the rent. Hence, it is very important to select tenants who can pay the rent on time and have a great credit score. This company provides the equivalent of a LinkedIn profile for renters. Hence, the landlord can pick a tenant who has a good credit score, which means he is able to pay the rent on time and has done so in the past. The tenant, on the other hand, improves and receives a good credit score by paying the rent on time. Furthermore, the tenant is able to opt for the deposit-free renting mode, through which the tenant provides a small fraction of the deposit and the rest of the deposit is insured by an insurance company. The firm has collaborations with the Direct Line Group (Insurance) and Experian and is backed up by Round Hill Capital (Property).

Company Name: Ziroom
Website: http://www.ziroom.com/
Target user: Landlords
What they do: This firm based in Beijing enables landlords to check the credit score and credit worthiness of a tenant before renting out the property.

Company Name: CrediFi
Website: https://www.credifi.com/
Target user: Commercial real estate investors and borrowers
What they do: This company provides a data and analytics platform for the commercial real estate market. Portfolio managers have access to easy-to-use search software that transparently filters search criteria to 200 billion data points to be able to find and analyze owner contact information for securitized loans from underlying commercial properties across the US and the corresponding balance sheets. Also, the borrower has access to over 10,000 lenders in the US and can filter the search criteria through regional activity, property type, loan size, and a bank's corresponding loan exposure. In order to find the right business partner effectively, a mathematical algorithm helps to find the best terms and conditions for both parties and decreases cost and time.

Innovative process automation firms

Company Name: Kofax
Website: https://www.kofax.com/
Target user: Banks and loan lenders
What they do: The company provides a software solution to digitize and automate processes that intersect from front office to back office operations. This will result in a reduction of human interaction with paper documents, which will lead to a reduction in errors by employees, reducing cost and time. The software will replace manual input of data by reading scanned documents.

Company Name: Finastra
Website: https://www.finastra.com/
Target user: Banks and loan lenders
What they do: This FinTech firm has acquired Mortgagebot, which offers a cloud-based lending operations software service. This end-to-end mortgage platform speeds up the entire mortgage lending process, from application and origination to processing and document preparation. Borrowers will receive immediate results, rates, fees, disclosures, and approvals. Additionally, it has a paperless processing solution, which means it contains a built-in digital imaging tool. This efficiently organizes, indexes, and sorts required documentation, allowing lenders and borrowers to easily track, share, and transfer required files.

Company Name: Rate Reset

Website: http://www.ratereset.com/

Target user: Mortgage companies, lenders, banks, and credit unions

What they do: This business-to-business company works with mortgage companies, banks, and credit unions and provides software that resets a loan. The technology behind the software enables mortgagers to use a feature of a loan to reset it. Without the obligation to refinance, this rate reset turnkey solution enables customers to reset the terms of their mortgage simply by using the lender's website to determine the new monthly payment, interest rate, and mortgage loan term. It can be executed simply by using the electronic signature function, without the requirement of further documents to be signed.

Company Name: Roostify

Website: https://www.roostify.com

Target user: Mortgagers and borrowers

What they do: This company provides a business-to-business solution by offering an online platform which has digitized the mortgage process for the end user. Its aim is to reduce time and cost and to increase transparency for all participants during the mortgage application process.

Company Name: Blend

Website: https://blend.com/

Target user: Banks, lenders, and financial institutions

What they do: This company focuses on banks and financial institutions lending money to the end-user. By improving the mobile online mortgage application process for the customer, they increase speed and efficiency. Mortgagers can authorize and link to several data providers. Through their guided workflows, connectivity, and automation, they have increased transparency and provided access to primary-source data.

Digital loan facilitators

Company Name: Habito
Website: https://www.habito.com/
Target user: Mortgagers and borrowers
What they do: This company is an online mortgage broker, which is using artificial intelligence to analyze the customer's data and the mortgage market in order to speed up the mortgage application process and find the best suitable mortgage for the customer. The company goes one step further by digitally brokering the mortgage.

Company Name: LendingTree
Website: https://www.lendingtree.com/
Target user: Borrowers and mortgagers
What they do: This company is a mortgage comparison platform with a large network of banks and lenders. It tries to research the best possible product for the borrower to provide mortgagers a number of proposals from numerous financiers within a few minutes. The platform offers access to money lenders proposing different types of property loans.

Company Name: Hypcloud GmbH
Website: https://www.hypcloud.de/en/
Target user: Construction companies and property builders
What they do: This company is a digital platform permitting property builders to negotiate senior-secured funding for commercial real estate with a number of lenders simultaneously. Due to its digital infrastructure, it can be an efficient way to obtain funds in only four steps. The algorithm matches the suitable bank and lender without inefficiently searching the entire market.

Company Name: RealAtom
Website: https://realatom.com/
Target user: Commercial real estate borrowers
What they do: By focusing only on the commercial real estate market, this real estate debt platform enables borrowers to find and close deals faster. Lenders, brokers, and borrowers are able to work on property investments in the range from $100,000 to more than $100 million.

Company Name: Trussle
Website: https://trussle.com/
Target user: Mortgagers and borrowers
What they do: This is one of the UK's first online brokers, founded in 2015. The platform focuses on the end user by only comparing loans related to real estate. Furthermore, it does not charge the borrower and has a simple, intuitive, and fast method to identify the best lender. It also supports the borrower with additional customer service.

Digital lenders and underwriters

Company Name: EstateGuru
Website: https://estateguru.co/
Target user: Real Estate lenders and borrowers
What they do: This firm provides a digital platform for firms with a track record in real estate deals who are looking to obtain funds directly from an investor. Investors are able to receive high returns. In the past, the platform has provided their investors a dividend of 12.5% and ensured no capital loss.

Company Name: Better Mortgage
Website: https://better.com/
Target user: End user borrower
What they do: This company is an online mortgage lender that has digitized every step in the financing process. It guides the borrower intuitively trough all parts of the application process and keeps all participants informed about the progress. The firm is supported by Kleiner Perkins and Goldman Sachs and was founded by former Google and Spotify employees.

Company Name: Guaranteed Rate
Website: https://www.guaranteedrate.com/
Target user: Mortgagers
What they do: This company is a large lender from the US founded in 1999, which introduced the world's first Digital Mortgage SM technology to reduce interest rates and mortgage application fees. The combination of technology and customer service has enabled the firm, which is licensed in all 50 states, to reach a market-leading position in this industry. It is the largest retail mortgage lender with approximately 210 offices.

Company Name: LendingHome
Website: https://www.lendinghome.com/
Target user: Mortgagers
What they do: This company is a fully digitized online mortgage provider. It also collaborates with real estate agents and mortgage brokers with the aim of delivering the best application experience for the borrower. The entire process becomes transparent, and the firm aims to close the entire mortgage process within four steps.

Company Name: Quicken Loans
Website: https://www.quickenloans.com/
Target user: Mortgagers and borrowers
What they do: This company is one of the largest—if not the largest—mortgage and property lenders in the US. As it is focusing only on home and property loans, it has a variety of tools and products for this market only. It is fully digitized, but supports the client with a customer service helpline in order to increase transparency speed and improve the process of the mortgage application. Furthermore, it is also gathering a rather huge database of customer financial information, which it markets.

Company Name: Sindeo
Website: https://www.sindeo.com/
Target user: Mortgagers and borrowers
What they do: This firm is a San Francisco-based mortgage marketplace, which provides the tools, information, guidance, and access to mortgage advisors in order to find the most suitable mortgage and lender. The aim is to provide recommendations by improving the mortgage application process and cost structure through technology and improved transparency.

Company Name: SoFi
Website: https://www.sofi.com/
Target user: Investors, mortgagers and borrowers
What they do: This digitized online lender was set up in April 2011 in San Francisco and has raised more than US$2 billion. It provides a variety of loans, including mortgages for residential properties. Its online platform has reduced cost, and improved efficiency and transparency. It is also combining a sophisticated online platform with over-the-phone customer service.

Company Name: WeLab
Website: https://www.welab.co/en
Target user: Banks, lenders, and financial institutions
What they do: Founded in 2013 in Hong Kong, this company is focusing on seamless mobile lending experiences and aims to democratize funds. It has developed a proprietary risk management technology to collect unstructured mobile big data to analyze and provide credit decisions for individual borrowers. China's and Hong Kong's leading mobile lending platforms, Wolaidai and WeLend, are owned by this firm. Conventional financial institutions are also using these FinTech-enabled solutions for their customers.

4

The Transaction

"Believe what you see and not
all you hear."

–Indian proverb

Real estate transaction

After looking at different ways to raise money in order to acquire a property, it is time to look at one of the most demanding phases of the property cycle, which is the real estate transaction.

It is the most stressful situation for the seller and also for the investor. All parties have an interest in completing the transaction, but there are a number of variables that can stop the process. It is an uncertain situation, where the other party or any other stakeholder might find a reason not to conclude the transaction. All stakeholders are fulfilling their duties, and when something does not seem to be right, the participant cannot proceed.

As every property is different, a thorough due diligence has to be done by all the stakeholders in the transaction. A very good broker can make a difference. He should be able to communicate clearly and precisely, and chase down the participants to ensure that they are delivering the service in a timely manner. He is able to bring transparency into the transaction process.

Without technical and digital solutions, this work is time-consuming and stressful for the broker. Hence, his work can become expensive.

The real estate industry has not provided a solution to this problem. This can be intentionally or unintentionally, but there are start-ups focusing on this issue by delivering digital solutions to increase transparency and speed during the transaction process. Hopefully, the benefit will be less stress and lower costs for all participants.

By analyzing firms supporting real estate sellers, it is possible to identify current problems in the property market and trends for start-ups to provide a better solution. By looking at the other perspective—the property investor's—it is also possible to see why a real estate transaction has been so expensive. By using methods and strategies that reduce cost through transparency and speed, it is possible to see where start-ups are trying to provide solutions for established real estate companies.

Finally, it is an exciting time for real estate because a local product becomes not only global, but also digital. A property can be sold and bought by linking participants who are not close to the property geographically. Digital solutions enable the participants to see the property through virtual reality, and sign contracts digitally to buy or rent out the investment property.

Selling real estate

Selling a property can be a very big decision. It can be assumed that everyone who decides for the sale has conducted in-depth due diligence before making the decision. In the past, it was very difficult to obtain up-to-date or real-time data and information about the market, and it was also very difficult to judge an accurate fair market value of the property. In a lot of countries, this information is not generally available and not transparent.

In some locations, like New York or London, it is now possible to obtain information about property prices and even future predictions. Companies have realized that obtaining this information is not only crucial for commercial real estate investors, but also for private property buyers. They have created a marketplace for real estate data. Past, current, and future property prices are available to participants in this marketplace, and the marketplace for data and analytics is growing fast.

A real estate agent who is providing information about his deals now can be rewarded by such a marketplace with access to potential future deals. In this way, a digital platform can provide very accurate real estate market predictions.

Once the price has been determined, it is time to market the property. The real estate market has come a long way from the days of advertising properties in newspapers and magazines to today's real estate market platforms that offer pictures, videos, community information, and virtual reality tours. A virtual reality tour of a property broadens awareness of its availability for the seller and broadens the selections at the fingertips of buyers—even worlds apart. This increases the number of potential buyers and enables sales without the need to conduct a physical viewing, which can be time consuming to schedule and conduct.

Once the decision to sell has been made and the price has been determined, it is now the job of the vendor to accumulate all the required documents the buyer will require. Previously, the seller would instruct his solicitor to obtain all the legal documents to provide to the selected buyer. With many different documents required at different times from different stakeholders, emails were usually being sent back and forth. In the commercial real estate world, a few years ago companies started to provide virtual data rooms (VDR) to store and share these documents. Furthermore, they are providing a way to ask questions about the property through the VDR, so questions and answers can be visible to all participants.

Some landlords today may decide to sell a property without a broker, leaving all the work a broker would do to be conducted by the owner. There now are digital tools to support this endeavor. For a small fee, the vendor can decide to use one of several digital platforms to execute most of the brokering work, and also provide a guideline and a path to follow in order to complete the transaction.

Many of these tasks were done in the past mainly by property brokers who had knowledge of the market, a large client database, expertise in marketing, and time to chase down all stakeholders and collect and provide documents as required. This added cost to the transaction. However, with the use of technology, the entire process of the transaction can become more efficient and effective. Technology can use data and algorithms to create the perfect match between an investor and a property. It can also create transparency to show the status of the sales process to the landlord and to the entire market competing to sell a property.

Finally, a number of hybrid models have been established, where a broker is using technology to complete a task such as advertisement, marketing, or the creation of the property's virtual reality tour and leaving other tasks to the vendor. This optimizes the use of expertise and reduces cost and time during the transaction process. Overall, the important factor is implementation of technology in all stages of the transaction.

The final strategy using big data involves firms buying a property from a vendor at a fair market price and selling it on. This can be very helpful for the vendor in order to avoid any stress. The company flipping the property has the expertise and knowledge to optimize before selling on.

Buying real estate

For the buyer of the property, it is crucial to conduct due diligence before making a buying decision. Among many factors, the characteristics of the property are relevant to an investment decision. As the commonly used phrase "location, location, location" indicates, it is important to obtain the information about the micro and macro location. Furthermore, it is relevant to understand the legal and financial situation, and the condition of the building. For the latter, the conventional method was to dispatch a surveyor or a technical due diligence expert to the building to inspect and report any defects which might reduce the price and value of the property. During a commercial real estate transaction, technical due diligence is very time consuming and can become very costly.

There are technologies now that use a drone to fly over the building and take pictures of the roof, the walls, the exterior of the building, and the land. These pictures can be utilized in the office to identify cracks, water leaks, and other defects on the exterior. With real estate market participants starting to enter information about transactions or the buildings, it is even possible to conduct a desktop review to determine capital expenditures and predict operational expenditures.

For the residential real estate market and for private homebuyers, technical innovation enables them to modify and determine a new layout for a new construction and the unit of the flat they are interested in buying. Furthermore, it is also possible to analyze and determine the amount of daylight a property is receiving due to its position.

Therefore, due diligence starts by obtaining data and matching it against the investment requirement. New start-ups are obtaining big data to use their algorithms to evaluate a property's location. This time-consuming task was made efficient and effective through deep learning and artificial intelligence and is being sold to real estate investors and asset managers. Using this data enables the creation of scenarios important to the investment decision. Digital platforms provide a way to use this interoperable provision of data and analytics to create standardized reports for the decision maker.

In order to speed up the transaction process, it is possible to use new digital platforms. The platforms are using information and data of the investors, brokers, and vendors in order to create the perfect match. The investor is not required to conduct lengthy purchases or to view properties that do not meet their criteria. Hence, they can target very effectively only those few properties that are a perfect match.

The digital way of involving other participants during a transaction process provides transparency and ensures completion deadlines can be met. For example, the status of completion of individual tasks can be seen during the transaction process by all participants. Hence, the broker does not need to personally track the work; instead this can be left to the software. By utilizing technology, it is possible to decrease the time spent on a transaction by the broker, which reduces costs.

Innovative Firms

In this chapter, we are looking at all elements and innovative companies that deal with the transaction's back-office work: the legal documents, the blueprints, the appraisals, the notary, the site survey, and all the true due-diligence work, including background checks and credit histories.

Virtual data room providers

Company Name: Intralinks
Website: https://www.intralinks.com
Company Name: iDeals Solutions Group
Website: https://www.idealsvdr.com/
Target user: Commercial property brokers, sellers, and buyers
What they do: During the commercial property and real estate portfolio transaction, a number of documents have to be made available to all participants and finally exchanged. Virtual data room (VDR) companies are providing platforms where the seller can upload all the documents for the buyer and the due diligence team to access. As these documents are confidential and large, the companies providing this digital platform also provide a number of security features, such watermarks and access tracking. Additionally, functions like a full text search, question and answer sections, bulk uploads and downloads, audits, and reports can be used, too. With the implementation of new technology such as artificial intelligence that can read documents, it is now even more efficient to search for the right documents.

Company Name: EVANA AG
Website: https://www.evana.de/
Company Name: Leverton
Website: https://www.leverton.ai/
Target user: Commercial property brokers, sellers, and buyers
What they do: During a transaction process, thousands of documents are changing hands within the commercial real estate industry. These documents will need to be analyzed by a number of different stakeholders and are sorted in a certain folder structure. A lot of time can be saved by automatically sorting the documents. So, a solicitor can access immediately legal documents, or a surveyor can access facility management documents without a long search. These two firms from Germany are leveraging artificial intelligence and transferring the analysis of the documents directly to the computer. This means that stakeholders during a stressful transaction process can search for the appropriate document, and even more deeply to find the appropriate content within the document.

Company Name: HouseCanary
Website: https://www.housecanary.com
Target user: Residential property brokers, sellers, and buyers
What they do: This technology-focused company offers a platform for residential real estate data and analytics that provides analytics, valuation reports, agile evaluation, and appraisals. By using machine-learning algorithms, it provides an automated valuation mode, which captures nearly all US residential homes. The automated and data-driven value report uses big data to determine the current value and predicts the value of the property for the next three years. If a condition report or a survey is required, a specialist will be able to conduct an inspection and write an agile evaluation report on top of the data-driven value report, which adds the condition of the building after an inspection has been conducted.

Virtual tours

Company Name: Habiteo
Website: https://pro.habiteo.com/
Target user: Brokers
What they do: This company from France enables project developers and real estate agents to sell a property using the latest technology for presentations. The real estate investor is able to go through the property using 3D virtual reality presentations to view a 3D rendering of the building yet to be constructed.

Company Name: Matterport
Website: https://matterport.com/
Target user: Brokers
What they do: This company provides solutions for property agents to create 3D presentations of the property. The real estate agent is able to use his 3D camera to take pictures of the property being marketed. Those pictures can then be uploaded to the company's digital platform, which then creates the virtual tour. The broker can use this 3D model in his marketing campaign.

Company Name: EyeSpy360
Website: http://www.eyespy360.com/
Company Name: KOMA Ltd.
Website: http://www.koma.guru/
Company Name: Giraffe360
Website: http://giraffe360.com/
Target user: Property sellers and buyers
What they do: A new technology that has emerged over the recent years is virtual reality. VR is not only useful during the construction process, but also at the end of the property lifecycle when a property has to be sold or rented out again. Potential tenants might not be able to go and view the property physically, but require additional information to imagine the property. Virtual

reality enables people to imagine how something can look in a virtual world in 3D, if companies are offering 360-degree virtual reality presentations. This means a landlord or property management company can make photos with their 360-degree camera and either upload it or provide it to the platform where the tenant is able to view the property. It is possible for the tenant to virtually go through the entire property and see it from every angle.

Company Name: NavVis
Website: https://www.navvis.com/
Company Name: Insider Navigation
Website: http://insidernavigation.com/
Target user: End user
What they do: Many people today are using Google maps or other GPS-driven tools to navigate, primarily outside on a street or highway. However, what are you going to do if you are inside of a large building and you need to go to a certain location within the building? These companies provide a solution that maps the interior of a large building and creates a 3D or augmented reality (AR) view to navigate through it.

Company Name: Spectando
Website: http://spectando.com/
Target user: Property sellers and buyers
What they do: This company enables the user to make a 360-degree virtual reality tour by using their app and without any further equipment. Their software extracts data from floor plans and pictures to create photorealistic 3D models and virtual reality tours.

Company Name: ALLVR GmbH
Website: https://allvr.net
Target user: Investors and agents
What they do: It is difficult for real estate professionals to communicate blueprints, plans, and documents to clients and other stakeholders who are not architects or designers and struggle to visualize how something is going to look after the project has been built. Hence, very often during the construction project, a showroom must be built as one of the first units to allow people to see what a unit looks like when completed. The company uses virtual reality to create a digital visualization of a project and property.

Company Name: realiz3d
Website: www.realiz3d.net
Target user: Investors and agents
What they do: This company enables the use of existing floor plans or creates floor plans to view a property in 2D or 3D, and also provides a virtual walkthrough of an incomplete or unfurnished building. This helps to understand how the building is going to look once it's completed and allows changes in the design before the actual construction.

Drone technology

Company Name: Spotscale
Website: http://spotscale.com/
Company Name: Skycatch
Website: https://www.skycatch.com/
Company Name: FairFleet360
Website: https://www.fairfleet360.com/
Target user: Architects, master planner, developers, and more
What they do: In order to replicate and visualize an environment in a digital format, it is now possible to capture pictures of the environment using drones. These companies provide a solution which captures, analyzes, and processes 3D models. Drones provide a competitive advantage in comparison to conventional methods of capturing environmental data and pictures. These three firms aid in data gathering, topographical mapping, landscape design, and runoff control, and, in essence, replace flyovers by manned aircraft or the use of expensive satellite imaging.

Company Name: Dronotec

Website: http://www.dronotec.com/

Target user: Commercial property surveyors, investors, and sellers

What they do: Ahead of a purchase, a buyer must conduct technical due diligence inspections and make technical surveys of a building to understand the condition of the property. In the past, firms visited the property physically to look for any damage, which was very time consuming, and not all of the exterior of all buildings was visible. Furthermore, it might have been difficult to reach certain locations, such as the rooftop of the building. Drones allow these inspections to be done more quickly, and thoroughly. This company flies the drone over the building to capture pictures, which can then be used identify potential defects such as cracks, leakages, et cetera. Furthermore, it is possible to review the building from different angles and create 3D renderings.

Sales process

Company Name: Houzeo
Website: https://www.houzeo.com/
Target user: Property owners
What they do: Property owners who decide to sell the properties by themselves without using a property agent can find themselves taking up a task that can be very time consuming and stressful. This company has created a digital platform that makes the decision to sell a property without a broker look easy. This digital for-sale-by-owner platform breaks down the processes into six steps. It supports listing and marketing, determination of the asking price, disclosures, support for the attorney with escrow, comparing offers, and the closing.

Company Name: SkenarioLabs
Website: https://skenariolabs.com/
Target user: Commercial property surveyors, investors, and sellers in need of technical due diligence
What they do: This company from Finland collects real estate data from the UK and the Scandinavian market to provide analytics to commercial real estate investors, managers, and funds. Those clients can rely on the services and reports which have been generated through AI usage and big data analysis. The technical survey of large stocks of real estate can be used to assess an investment and predict the future risks of the buildings. It is also possible to determine the fair value of the commercial and residential real estate, as the firm is using a standardized algorithm to determine the fair value of the property.

Company Name: Geospin GmbH
Website: http://www.geospin.de/en/
Target user: Commercial property investors and sellers
What they do: By using spatial intelligence, it is possible to optimize investments into the ideal location and infrastructure. This firm uses spatial data, demographics, infrastructure, weather, and even social media information in order to determine the best spatial strategy and location of real estate investment. They combine Geo deep learning, big data, and predictive analytics in their algorithms.

Company Name: HABX
Website: https://www.habx.fr/
Target user: Residential property buyers, architects, and developers
What they do: This company enables the buyer of a new development to modify the residential unit they are interested in buying. The platform is digital and easy to use for non-professionals. The final draft of the chosen change shows the additional cost of the modification, which then can be approved by the architect.

Company Name: Solen
Website: https://www.solen.co/
Target user: Brokers, buyers, sellers, and architects
What they do: The quality of the interior of a property can also be determined by hours and quality of light within a building. This company from France certifies the number of hours of direct sunshine per day, the room luminosity, and strength of the light on the property. Brokers and developers can use the service to evaluate their properties. The digital platform enables the use of an app which predicts the trajectory of light through augmented reality and calculates with an algorithm the amount of light within building.

Financial due diligence & data analytics

Company Name: CompStak
Website: https://compstak.com/
Company Name: RealMassive
Website: https://www.realmassive.com/
Company Name: Real Capital Analytics
Website: https://www.rcanalytics.com/
Target user: Commercial property brokers, sellers, and buyers
What they do: During a sales process, it is very important to determine the correct selling price. If the list price is too high, the vendor might sit on his property for a long time before he is able to sell. If the price is too low, he might lose potential income. These companies provide a platform that delivers data and analysis of the real estate market. Market participants, such as property investors and real estate agents, register information from their transactions on this platform and can draw comparative data from it. Hence, this platform grows in data and can make in-depth analysis and predictions about the local real estate market.

Company Name: uDA. urban Data Analytics
Website: http://www.urbandataanalytics.com/
Target user: Investors
What they do: This data analytics company from Spain focuses on the Spanish and South American real estate markets and uses an algorithm and big data to calculate the fair market value of properties.

Company Name: DataScience Service
Website: https://ds-s.at/
Target user: Investors, owners, and sellers
What they do: This company from Austria provides data analysis and big data analytics for the real estate investment and banking sector. They collect, mine, use an algorithm, prepare, plan, and predict the behavior of the real estate market, leaseholders, and investors by using big data. Furthermore, they create an analysis, and aim to predict when a leaseholder is going to cancel a tenancy agreement.

Company Name: REalyse
Website: https://realyse.com/
Target user: Investors
What they do: This company collects big data from the residential property market in the UK to make it easily accessible for developers, investors, lenders, and consultants. They're able to search through different datasets to compare and find comparative ratios for investment analysis. Historical data can be obtained to create scenario analyses and reports after exporting the data.

Company Name: Mashvisor
Website: https://www.mashvisor.com/
Target user: Investors
What they do: This firm's platform provides neighborhood information and analytics such as cash flow, ROI, cap rate, voids, and rental market ratio indicators to identify secure investment market potentials for traditional and short-term lodging (Airbnb) properties.

Digital notary

Company Name: MyNotary
Website: https://www.mynotary.fr/
Target user: Notaries and lawyers
What they do: This French company has digitized the work and processes of the notary during a transaction. The participants of the transaction can access the platform to submit documents and sign paperwork as requested.

5

Getting Smarter

"If you plan for one year, plant rice. If you plan for ten years, plant trees. If you plan for 100 years, educate mankind."

–Chinese Proverb

Today, technology also is taking aim at the long-confusing and mysterious art of caring for or managing a property, be that a home, an office building, a condo, or an incubator. Much of this stems from advancements of all types that are moving rapidly toward smart houses with all mechanicals and operating systems chock-full of sensors and communications capabilities linking them not only together, but to remote locations featuring sophisticated monitoring, analysis, and reporting software.

In the home, this allows for close monitoring of heating and cooling, timing of lights, and things like the smart doorbell. In the office, gathering data about climate, motion, sound, and other usage trends allows for smart systems to optimize efficiency, control costs, and even regulate parking lot usage. Others optimize design, layout, navigation, and so on—all reporting back to the owner or manager of the property. This allows them to optimize routine maintenance like cleaning, anticipate repairs, and promptly respond to outages or breakdowns, and report all of this in real time to investors or shareholders.

Asset management, property management, and facility management

In this part of the book, we will discuss the management and maintenance of a property, starting with asset management, which is focused on commercial properties. We will compare the old way of working with asset management and compare it with the new technologies that have already entered the market in order to reduce the workload of the asset manager. We will look at firms providing transparency and improved communication methods during the operational process.

The main aim of real estate asset management firms is to increase and optimize the value of the building and to manage and maximize the investment returns. Throughout the entire lifecycle, asset managers anticipate potential risk by analyzing data and forecasting profitability for the owners.

The investors or owners, who can be small family offices, large insurance companies, or pension funds, hire asset managers to ensure that vacancies, arrears, compliance issues, repairs, and maintenance are at optimal levels, while rental income is at highest levels. By analyzing the market, data, and reports, they are able to implement the best possible strategy for the performance of the related portfolio. Often the asset manager hires a property manager to conduct daily operational tasks on individual properties, while he focuses on financial matters. This also means deploying the capital of the investor to generate the expected return. Hence, the asset manager either buys a portfolio of buildings or builds up a portfolio of assets.

A bit later, we will look at property managers and their daily work within the commercial properties, and also at private landlords. We will evaluate how technology is improving communication and transparency for administration tasks. We will then look into technical and digital ways to improve the leasing process from the landlord's view and the renting process from the tenant's view.

Finally, we will look into facility management and all the changes that are coming through the installation of sensors in buildings, which are becoming increasingly smart.

Asset management

Overview & problems

As diversification plays an important part in risk management, asset managers have to mitigate risk using different real estate asset classes or geographical asset distribution.

This could mean they cannot be close to their properties all the time and must rely on the local property manager or facility manager. Therefore, it is crucial that transparency is provided, which means the owner or asset manager knows the status of their buildings, any upcoming and completed repairs, and potential changes of tenants. This was mainly done through property managers in the past, but now can be accomplished through technology. An asset manager can draw data from sensors in the building that give them information about their technical status. Digital reporting systems allow every task accomplished by the property manager to be transparently visible to all stakeholders.

Property types can be differentiated into the common ones: office, industrial and logistics, retail, and residential. The following assets are emerging as property types from a niche market: hospitality and smart hotels, self-storage, student housing, and senior housing.

As asset managers generally work on funds, which have a certain strategy, this can be divided into core, core-plus, value-add, and opportunistic.

Core:

Properties within this category are low risk, but also have a low potential return. A manager has to ensure that the property will remain stable and fully leased. Assets are typically located in Triple-A locations, which are mostly in gateway cities. It can be stable residential or office properties with long-term leases.

Asset managers within these categories require a lot of data to analyze in order to make a buying decision. The data can be related to the entire country, political changes, GDP, etc. (macro); and location-specific, such as prices of the building, the street, district, etc. (micro) that can be used to gather enough information to predict real estate prices in the future and the potential for change. The reports, information, and data should be easily accessible even for investors, who are located at different geographical locations.

Core plus:

Properties within this category are similar to core assets, but are leveraged by 20% to 30%. Information about the tenant market can be very relevant within this investment class. Asset managers should be able to work with property managers and facility managers, who will have access to digital marketplaces in order to ensure full occupancy rates. Insights into the building and its performance can enhance the profitability. This means it is necessary to understand tenant requirements and make repairs and address maintenance issues quickly.

Value-add:

Asset managers within this category take on portfolios or properties that require improvements either physically, in management, or operationally, and some suffer from capital constraints. Properties within a value-add strategy are leveraged between 40% and 60% and face a medium-to-high risk of losses if the improvement plan cannot be implemented. However, it also provides a medium-to-high return on investment. The different managers need to work collaboratively within this category and require an in-depth understanding of the micro situation, which means an insight into the building and defects in order to improve the value of the property. Implementation of sensors or other data acquisition strategies, such as the measurement of wi-fi strengths, brings visibility and transparency into the utilization of space. Furthermore, the installation of sensors in facilities like the elevator, the valve of a water supply, or windows can predict maintenance requirements. The building automation system within the property can decrease utility costs and increase productivity and profitability. Furthermore, it is also possible to create additional services outside the building and provide users or occupants with facilities to be accessed via the smartphone. This could be an app which lets the tenants use services in the building or provided by neighboring service providers, and also let the users of the building communicate with each other or the operators. A community can be created, which enhances the experience of the end user.

Opportunistic:

Real estate within this category can be considered as high-risk assets, which can deliver a high return after the manager has implemented his improvement strategy. This can mean an investment in a property that is run down or has major problems, such as asbestos. The asset manager takes on this challenge, removes the asbestos from the building, and refurbishes the building in order to sell it for a higher price. An opportunistic strategy can also mean the management of the development project. The level of leverage can be 60% or higher.

Asset managers working on properties under this strategy are now able to use Building Information Modelling (BIM) tools. This means that, before the entire refurbishment work begins, they are able to plan, see, and simulate the entire process and visualize the construction digitally before they start issuing the work to the contractors. By understanding how and when the individual components and facilities are installed, it is possible to plan refurbishment processes or optimize energy consumption within a building. Without a 4-dimensional model, errors are visible just after the installation of the components have been completed. Hence, it is useful to run the installation process digitally, before actually doing the construction work.

The asset manager's performance

The asset manager is not simply paid at a fixed rate, but also has the owners and investors looking at his performance. This means, the higher the value of the assets under management, the higher will be the fees the asset manager is able to charge. Hence, the asset manager is incentivized to increase the value of the building and the performance of the portfolio by reducing his own cost and time. This can be different for the property or facility manager, who might not receive a bonus according to the performance of the building, being paid according to the completion of a task at a fixed rate or on an hourly basis.

In a nutshell, asset managers in the past were forwarding the daily operational tasks and heavily relying on the property manager and the facility manager. Consequently, a good asset manager was the one who selected the best property manager and the best facility manager.

Interoperable systems increase efficiency

New technology now makes it possible for the asset manager to be virtually closer to the building and have much more insight using data. An asset manager who needs to forecast rental income, capital expenditure (CAPEX), and operational expenditure (OPEX) previously had to rely on general market statistics from the past to predict the future. With the use of sensors, it is possible to obtain real-time data for the building. This allows an asset manager to determine, for example, the consumption of energy by a specific floor, room, or even tenant. Alternatively, it can measure the longevity of an elevator, which might require replacement in the next two years. This data is then accessible from everywhere in the world at any time, and the asset manager can create a budget based on his forecast.

Furthermore, the asset manager must transparently show the owner or investor all the details of the property and portfolio. Technology enables the use of information from the property manager's system, which is compatible with the system from the asset manager. Hence, different digital systems are now interoperable.

Efficiency can be created for true scalability, which means the asset manager can use one system and one standardized algorithm and apply it to different buildings to receive the same kind of reporting, if the same tasks and strategies have been standardized and implemented once a call-to-action scenario is activated. This could be applied to any changes, such as a tenant moving out and a new tenant moving in.

Some asset managers and portfolio managers treated properties as securities and products of the financial market. This meant that they were not familiar with the requirements of tenants, defects, or problems, and could not learn from and use the feedback from the building and the tenant. Hence, the building did not evolve or improve and tenant requirements were not met. With the usage of technology and by obtaining data to analyze and evaluate, it is now possible for the asset manager to directly understand the accurate and real-time usage of space, facilities, and time.

The asset manager does not need to go to the facility or property manager to ask which improvements need to be done. Instead, through the interoperability of different software, the asset manager can see the data from the building and from the tenant directly. An example is the use of heating in office buildings. Sensors can measure the heat or noise level in order to show where employees are located. It was impossible to understand as accurately as it is now which areas of the office were empty and which were maybe even overcrowded.

Office managers now can change the layout of the office to use all space equally. This data can also show which area has been used to determine which area has to be cleaned and which does not need cleaning. This can maximize utilization of the office space, increase productivity of the employees, and decrease costs for things like cleaning.

The property owner or the asset manager is required to find tenants and always targets full occupancy. It is now possible to use online marketplaces to find potential tenants who are not in a database of an agent. This also means that there are no geographical boundaries or language barriers, as owners are able to advertise via virtual reality platforms to find international tenants in search of a new premise overseas. Then they should be able to sign a contract digitally.

Property management

One important business partner of an owner of a building is the tenant who is paying the rent, which generates the revenue. An empty building is merely a speculative investment without cash flow and without real creation of value. A property manager stands between the tenant and the owner most of the time, finding new tenants, dealing with the daily needs of the tenant, and paying the landlord the collected rent. The property manager reduces the workload for the landlord and, due to the specialization in this field, is able to scale and leverage the knowledge they have.

There are well-trained and professional property managers who are a blessing to the landlord, but there also are property managers who do not deliver what they promise. Property managers may be under very high pressure to accept more work than they can actually handle, or the demands of tenants may increase. Hence, the attractiveness of the lease for the tenant decreases for a poorly managed property.

Therefore, it is crucial to use technology to enhance efficiency, transparency, and communication. This will lead either to lower costs or lessened workload. The repetitive work of the property manager can be standardized by using modern SaaS solutions. The way a tenant communicates with the property manager has been changing from an analog way, which meant accepting a phone call, or sending a letter or fax, to a digital process. Digital communication allows the tenant to report issues with evidence (photos) via an app on a smart phone, via the Internet. Or systems may allow the building to detect and report an incident through sensors.

The demands of tenants are also increasing. The traditional tenant tended to be happy with a building they were able to furnish and maintain. However, the trend today is moving towards flexibility and short-term lets. Tenants rent a place as a service, like co-working or co-living space, without the hassle of furnishing and maintaining the building, cleaning the premises, and worrying about utility costs, taxes, or insurance.

So, tenants may prefer to pay a small premium for a rental package to a service provider like a property manager, who takes care of all these things and gives him a flexible tenancy agreement. The tenant can cancel and move on if required. Co-working and co-living providers have understood the change of tenant demand and are in high demand at the moment.

Co-working and co-living spaces create a new type of community within the building. This building connects people with different abilities and different demands, which can provide a productive synergy. Companies like Starcity, Common, and WeWork have created productive and flexible environments and a communities.

However, such flexibility in a location, with tenants paying different rates for different services, can cause problems, if the usage is not separated. For example, one tenant might book a package with flat rate access to a kitchen with a coffee machine and a meeting room. Alternatively, a tenant of a residential property may want to use the swimming pool one time only.

Luckily, technology can track tenants, their behavior, and invoice for usage only. Smart locks can give access to the meeting room and the kitchen with a coffee machine only to the paying customer, and can invoice the user of the swimming pool for one-time usage only. The flexible package tenant can access these rooms with their smartphones, too, with the added charges deducted via an app from their bank account.

We show how these new strategies are making property management more efficient and speeding up the system. In fact, by making the credit check/leasing process efficient, owners no longer need to worry so much about vacancies and they can be confident in quickly re-leasing. This also allows property owners to confidently enter into short-term leasing. Later, we will look at start-ups that are providing digital solutions for residential, commercial, and individual properties, for both landlords and tenants.

Renting out a property (landlord's view)

Renting out a property was a very stressful challenge. Within the individual residential industry, landlords decided to rent out the property by themselves by using pages like Zoopla or Rightmove in the UK, and in Germany, Immobilienscout or Immonet. These services helped to reduce empty properties, but now it is also possible to improve the process of renting out the property as firms offer a contract that can be signed digitally. This means a landlord living in a geographically different location is able to rent out his property much easier.

It is also possible to do a tenant credit check as an individual owner of a small property. In many of these instances, software packages, apps, or cloud-based SaaS are allowing people to do these things without hiring experts or making a contact that was needed before. A small property owner now can do this in-house, rather than paying an agent or site manager. Even if the tenant is not providing enough security, it is possible to obtain a rent guarantee from the property manager checking the creditworthiness of the tenant by using big data. A company can lease the property from the landlord and rent it out for higher rent and make profit on the margin. Nevertheless, if the landlord decides to rent out to different individuals, all of them have to make sure that their utility bills are paid. There are digital solutions where all tenants of a flat can view and pay outstanding utility bills via an app.

Within the commercial real estate market, property owners decided to pass on the task of letting to property managers or brokers. Tenants in search of new premises also contacted them, because they knew that they had a large database of properties. The intermediaries had a large database of potential clients, which they did not share. Hence, there was not enough transparency for the owner or for the tenant to avoid the intermediary.

New digital online marketplace platforms now provide this transparency and can bring the tenant and the owner together at a cost that is lower than using a manager or agent, in many cases also providing broader reach on both ends. It is now possible to rent out retail or logistic spaces on a short-term basis. The residential and hospitality sectors are merging, and both are providing furnished apartments that provide amenities such as furniture or a concierge, like a hotel.

New digital marketplaces make it possible to rent those furnished apartments on a monthly basis, focusing on business travelers who have work to do in a different location, but do not want to stay in a hotel for months.

Digitized marketplaces are the best platforms to use for landlords, property managers, and tenants. These marketplaces not only provide data and information that lead to the perfect match, they also can go one step further by allowing potential renters to view the place digitally through virtual reality. This increases reach to customers beyond geographic limitations.

Furthermore, co-working and co-living service providers take on the risk of vacancies and lease an entire building from the owner in order to rent out small and packaged workspaces or living spaces with a higher margin. The extra effort of creating a community and analyzing optimal usage of space and facilities pays off. These communities within the building and also the environment around the building are digitized to grant access to services and facilities using controls in an app. A community interacting with the app also can report any defects or needed actions (supply replenishment, cleaning, non-routine maintenance) that has to be taken immediately. The property manager can respond immediately with a contractor hire or product invoice seen by the asset manager.

Leasing a property (tenant's view)

Market experts have always claimed that real estate is local. However, it is digitally evolving through new technologies like virtual reality, augmented reality, and the Internet of things, which makes it possible to overcome geographic restrictions.

For a client needing to expand logistics, retail, or office space, a middleman such as a local broker or agent was previously needed to find the new premises for the client. This middleman role now has become digital, more transparent, global, faster, and cheaper. The middlemen are now the new digital marketplaces or platforms providing properties and directly connecting with clients.

Tenants can find new properties online and are not bound by their physical location. It is not only possible to see some pictures of a property, but also to access all information required to make a decision on whether to rent or not to rent. A number of them have already started to use the latest technology and provide virtual reality viewings online. The scalability makes it possible to charge only a small margin for bringing the two sides together, with profit coming from client volume.

This is not only applicable to the main markets, such as residential or office properties. New niche markets are emerging, such as for logistics, short-term retail, or even new types of retail premises, such as modified containers on a mountain or at the beach. Furthermore, a new niche has resulted from a combination of the traditional asset class, but used on a shorter contract basis. This means a tenant can stay in a furnished apartment for only one or two months and can rent a furnished co-working space for the same period. These new marketplaces, which are available online, provide access to this type of property.

Facility management & smart buildings

Facility management

The facility management sector is one of the fastest growing and changing environments within the real estate world, due to the technology emerging in the Internet of things (IoT). IoT devices and sensors enable cities and buildings to become smart. A smart building not only collects and provides data from its facilities, it has an automated system that enables different facilities to communicate with each other. The greatest improvement is the advancement of the sensor technology. This technology is even present in everyday devices, such as simple smart phones, headsets, carbon monoxide detectors, thermostats, smart watches, and cars. With the ability to transfer data and communicate, it is now possible to utilize the information gathered by the sensor and communicate back to take actions. For example, a simple electric thermostat senses an increase in temperature, which transfers this information to a central system, which activates the air conditioner. Those sensors are often not even bigger than a small microchip, and can be integrated in building components or products separately and simply installed on the wall, in a component, at a desk, or even a chair.

In the past, facility managers were responsible for repairs and maintenance and a large portion of the work was reactive. This means facility managers informed the property manager that a building might require some repair work and waited for confirmation to commission the work. Preventive maintenance was hardly possible due to the waiting time until repair work was approved. Due to such a long waiting time, facilities or components of a building could not be repaired before they completely broke down. This caused higher cost and consequently a lower return for the asset manager.

By using sensors and digital platforms that provide information about the condition of the building and its components, it is possible for the asset manager and the property manager to see data updating the condition. This means that, before the facility manager reports something, the property manager and the asset manager already have this information on the screen. Furthermore, it is not a subjective opinion of the facility manager or the contractor, instead it is an objective capture of the situation and condition of the building. It is therefore possible to react immediately and approve funds for maintenance work before the facility or system breaks down completely. The cost for maintenance can be much lower than an entire replacement.

Repairs and maintenance

While managing an entire portfolio, it is now possible to compare buildings and their expected ongoing cost. Digital platforms using big data are able to predict the operational expenditures and capital expenditure requirements of individual buildings. This was not that easy in the past, as cost that had accrued was generalized into one category, such as repairs and maintenance. With this new technology, it is now possible to determine where, when, and how a defect happened. With this information, it is possible to analyze the cause and install a preventive mechanism to avoid further defects.

For the development of new buildings, Building Information Modelling (BIM) should be used and the information should be passed on to the facility manager. He is able to use the data for the day-to-day operations. This means ongoing assessments of a number of components which are not visible without destructive measurements. For instance, it can be difficult to find the leak of a valve installed in between walls. Furthermore, it can be also very difficult to find the valve itself. With a 3D model of the building, it is possible to direct the contractor to the exact location of the component that caused the defect in order to fix it.

Sensors and Internet of Things (IoT)

Even if the building doesn't have a three-dimensional model, it is still possible to use sensors to optimize the usage of the facilities of the building. Sensors can be installed in a number of devices, such as windows, lights, walls, desks, chairs, et cetera. The application of IoT capabilities is endless. The sensors give a command to execute, if a condition has been met. The combination of different sensors interacting with each other in a building can already be considered as a property with an integrated building management system (BMS). It is a controlling system, which regulates the HVAC, lighting, and fire and security system of the building independently. This could mean that you are walking into a room and the light goes on because the sensor detects movement in the room.

Some companies are focusing on combining components to provide a solution for a certain problem. Other companies are focusing solely on one individual component of the facility of the building. An example could be a firm optimizing the usage and maintenance of elevators only. Another firm is focusing only on the sensors in a window, which detects temperature, humidity, and light in order to control the HVAC system or blinds. And there are companies focusing just on the HVAC system to optimize energy consumption. There are also firms with the sole focus of the air quality within a building.

Smart buildings

However, sensors do not necessarily automatically add up to a smart building. Buildings become really smart once they get a brain. This means the individual components, facilities, and the corresponding sensors need to talk and communicate with each other and are interoperable. For example, this could mean that if someone walks into a meeting room, the lighting, the projector, and heating or cooling goes on and informs other facilities in the building that this meeting room is now occupied. It could then inform the waiting client at reception that his meeting can start and direct him to the meeting room with indoor navigation through augmented reality, where the employee of the firm is already waiting for him. At the same time, it informs the other employees of the company that this particular meeting room is occupied. The Cube in Berlin is a building which will have those components communicating with each other.

Furthermore, there are companies that are not using sensors at all. Instead they create data to analyze and optimize the building by using given information, such as the energy bill or the given infrastructure, such as the strengths of the wi-fi signal. Firms which are able to utilize the data are the key to the future, as machine learning is able to automate and optimize the utilization.

Smart home

In the private and residential property market, a number of large firms such as Amazon or Google are already positioned to make the home smart. The focus for Amazon can be the ideal delivery, which means the courier will have an option to leave packages at the premise of the client even without the presence of the customer. This could be achieved by using smart locks. This large and growing market enables the resident to control access to a property without geographical barriers. If a courier is delivering a parcel or a good friend makes a visit, but the resident is not at home, the resident is alerted to the ringing doorbell on a smart phone through an app, sees the person via the doorbell camera, and is able to open the door or gate for the visitor.

On the other hand, Google might be interested in collecting resident behavior data and want to use it. Once they have this information they can analyze it and optimize advertisement. Firms like Google will receive the data the home is generating in exchange for a better and cheaper user experience, e.g. by managing home appliances remotely.

Optimizing energy consumption is also key for a smart home. By installing smart meters, it is now possible for companies to analyze the usage of energy of individual appliances and to predict which of those appliances might fail in the near future. This information also can then be forwarded to insurance companies to evaluate the premium according to the usage behavior of the resident.

Innovative Firms

Data analytics

Company Name: VTS
Website: http://www.vts.com/
Target user: Asset managers and portfolio managers
What they do: This company provides the leading software platform for the real estate asset management and portfolio management market. In the commercial real estate market, it works with the largest companies, enabling them to collect, analyze, and evaluate big data, which provides a competitive advantage. Brokers can optimize leasing activities by attracting the best tenants due to sophisticated data that now reveals information that wasn't readily available in the past. This means a sophisticated algorithm matches properties with tenants according to specifications and current information, instead of using past information. Portfolio managers can act on market trends, create sophisticated reporting, or find new deals. Asset managers can act on events and improve tenant relationship management. Due to the standardization of reports and processes, it is possible to reduce time and complete tasks faster. Furthermore, it is possible to optimize the real estate asset management operations and processes by integrating all the work into this platform, instead of sending out several emails to all participants during a transaction. The firm recently merged with Hightower, which is also providing software solutions for the commercial real estate market.

Company Name: Moderan
Website: http://www.moderansolutions.com
Target user: Small- and medium-sized real estate asset managers
What they do: This company provides an innovative solution for small- and medium-sized real estate asset managers. Their

solution provides a digital platform to record and report the performance of assets and conduct property lease management. The lean approach can work in conjunction with Excel or even entirely replace it. It captures data for the portfolio's key performance metrics, including income per square meter and total rental income.

Company Name: Assetti
Website: https://assetti.pro/
Target user: Small- and medium-sized real estate asset managers
What they do: This company provides a digital cloud-based asset management real estate software package. It provides features for portfolio management, asset management, property strategy, asset valuation, financial performance, property transaction, and tenant relationship management. It can be used to manage and receive property insight, monitor and reflect operations, provide instant access to information about the asset, and improve the property and asset management processes due to better communication and interaction with stakeholders.

Company Name: Exquance
Website: https://exquance.com/
Target user: Investors and asset managers
What they do: In order to make a buying decision on commercial property, in-depth financial due diligence must be completed and presented to the board of directors. Analysts have to accumulate the data and input them into their system, while asset managers have to review the performance of the units, and investors and the board of directors have to review the report. This company provides software that combines all investment analysis tasks into one system. It is possible to integrate a spreadsheet model into the system, create cash-flow models, and add legal structure models in order to track performance and generate reports.

Company Name: RealPage
Website: https://www.realpage.com/
Target user: Property managers
What they do: This firm is a leading provider of software and data analytics for the commercial property market. It provides a digital platform for real estate stakeholders to increase operating performance and returns. The digital cloud-based solution provides analytics tools for property management, sales and marketing, applicant screening, revenue management, spending, utility management, et cetera.

Company Name: TOWER360 GmbH
Website: http://www.tower360.co/
Target user: Asset managers
What they do: This company from Germany provides transparent, real-time insights on properties the asset manager or landlord is operating. It is a cloud-based platform that can be accessed via the web or by smart phone in order to view and operate asset management, leasing, and marketing tasks for commercial property. The aim is to provide higher efficiency and the interoperability of different data sources, such as CRM and Excel, and bring it into one platform. Hence, the user is able to mitigate risk, increase rental income, and reduce arrears and vacancies.

Company Name: EMonitor
Website: http://emonitor.ch/
Target user: Property management firms, housing associations, owners, and investors
What they do: A high occupancy rate is a very important factor during the asset management and property management process. This company provides a solution for property management firms, housing associations, owners, and investors. It has digitized the entire application process for renting out a flat and re-letting. It ensures that there are no more incomplete applications from potential tenants. It analyzes and provides recommendations to tenants about which properties might fit the best. This means it matches the existing and upcoming vacant flats with potential tenants. It also checks the credit scores of potential tenants. Due to the usage of spatial planning, and machine and deep learning algorithms, the owner of the property has the ability to monitor, analyze, and forecast vacancy rates.

Facility management

Company Name: Nutonian
Website: https://www.nutonian.com/
Target user: Facility managers
What they do: After completing a development project with Building Information Modelling, it is important to pass on the information to the facility manager. To ensure that the facility manager can continue to work on provided data, he needs a data analytics platform. This company provides a digital solution, so that the information which has been created can be utilized to optimize the performance of the building. The platform provides data science automation, which utilizes the data you have and the information you can act on. Data can be used to analyze, interpret, and strategize the ideal intensity of usage of certain components like the Heating Ventilation Air Conditioning (HVAC) system to optimize the usage.

Company Name: Deepki
Website: https://www.deepki.com/
Company Name: FirstFuel
Website: http://www.firstfuel.com/
Target user: Property managers and facility managers
What they do: These companies are focusing on exploiting data given through energy bills and invoices. It collects the data, analyzes it, and forecasts through an algorithm-optimized energy management in order to reduce energy, heating, or cooling costs.

Company Name: Placense
Website: https://www.placense.com/
Target user: Retail, hospitality, logistics tenants, owners, and property managers
What they do: Tracking foot traffic is relevant for retail, hospitality, logistics, transportation, and other industries. This company from Israel obtains such data without the installation of sensors to provide accurate and real-time information. A customer generates data through different apps on his smartphone, which is collected to aggregate insightful analysis of the customer.

Company Name: Distech Controls
Website: http://www.distech-controls.com/
Company Name: Entic
Website: http://www.entic.com/
Target user: Facility managers and property managers
What they do: These companies provide data analytics and digital building automation systems to reduce energy consumption. The aim is to provide the user a fully automated energy management system and let the building take control of its systems according to the usage of the occupier.

Company Name: WiredScore
Website: https://wiredscore.com/
Target user: Owners, asset managers, facility managers, and property managers
What they do: As the speed of the Internet service and connectivity is one of the most important factors in selecting an office space, this company has decided to rate office buildings and provide certification for their digital infrastructure. The firm not only certifies existing buildings but also development projects.

Smart Buildings

Company Name: Enlighted
Website: http://www.enlightedinc.com/
Target user: Developers, contractors, and facility managers
What they do: This company from the United States is a leading provider of sensors for smart buildings, focusing on providing Internet of things systems, applications, and solutions utilizing smart sensors. The smart sensors obtain data that can be used to optimize the lighting, usage of the room and space, and energy consumption. They have been successful in reducing costs for lighting, HVAC, space, and utilization, and improved productivity and safety.

Company Name: Disruptive Technologies Research AS
Website: https://www.disruptive-technologies.com
Target user: Facility managers
What they do: This company, which has its roots in Norway, provides solutions for the IoT industry and transforms a normal property to a smart building using sensors. It enables facility managers to monitor the condition of the units to improve performance of the asset. The sensors can write data to adjust temperature or reduce unplanned maintenance and repairs, which leads to overall optimization of the building condition. Sensors can also track the use of the office space, so it is possible to optimize the layout, floor plan, and furniture most efficiently. Finally, the data gathered by the sensors tracking the lighting, heating, and air conditioning can be used to analyze the most efficient condition to optimize energy consumption. The focus is on smart cleaning, comfort monitoring, temperature heat maps, preventive maintenance, HVAC control, energy savings, and workplace utilization.

Company Name: Greenbird Vertriebs Gmbh
Website: https://cleanbird.fm/
Target user: Facility managers, property managers, and tenants
What they do: This Austrian company uses sensors to detect the use of space and has an algorithm to calculate the intensity of use, allowing it to determine when cleaning is required. Furthermore, it takes into consideration the weather, or the absence of employees during holiday seasons, and promises to reduce cleaning cost by 15% to 35%. By creating transparency of usage, only truly required cleaning will be conducted.

Company Name: VISN
Website: https://www.visn.io/en
Target user: Facility managers, property managers, and tenants
What they do: This company makes a building smart by checking the real-time availability of rooms, providing optimum use of flexible working space, and adds a layout analysis. Furthermore, it provides smart cleaning and catering services, and facility managers can optimize energy consumption. With the app or the web service, employees are able to view and find available meeting rooms.

Work space analytics

Company Name: Lone Rooftop
Website: http://lonerooftop.com/
Target user: Facility managers, property managers, and tenants
What they do: This company provides a real-time occupancy platform to maximize and optimize use of the office. The cloud-based solution has an algorithm to analyze real-time occupancy data by utilizing sensors, the position of the employees in the building, and the use of wi-fi and the strength of the signal.

Company Name: Locatee
Website: https://www.locatee.ch/en/
Target user: Facility managers, property managers, and tenants
What they do: This company provides a digital platform to measure and analyze workspace requirements and utilization. It obtains data to measure the usage of individual desks and their availability to predict the highest productivity level. By using machine learning and indoor positioning algorithms, it is possible to obtain real-time IT infrastructure data, such as LAN or wi-fi usage. The platform collects, processes, and visualizes the data automatically. Using the data provided, the user is able to use workspace efficiently and enhance the employee experience.

Company Name: Toposens
Website: https://toposens.com/smart-building/
Target user: Facility managers, property managers, and tenants
What they do: This company uses 3D ultrasound sensors to ensure privacy by tracking employees in an office building. With specific ultrasonic signals, no personal information nor spoken language can be identified. With the principle of echolocation, the data on the flow of people through space and time can be utilized to analyze the use of offices, meeting rooms, retail stores, public transport, and any other spaces. This passive tracking system

makes it possible to optimize strategy and the usage of space anonymously.

Company Name: Sensorberg GmbH
Website: https://www.sensorberg.com/
Target user: Facility managers, property managers, and tenants
What they do: This company provides a digital and technological solution for the tenant of a co-working space to control their office space with a smart phone. The aim is to digitize a building with the usage of sensors so the tenant can interact with the building without the support of the office staff. This solution can also be implemented in the hospitality industry. Here, the guest can check in alone and interact with the building without the support of the concierge or receptionist.

Company Name: Spaceti
Website: https://spaceti.com/
Target user: Facility managers, property managers, and tenants
What they do: This company provides a one-stop solution by digitizing an office building using sensors to collect occupancy, environmental, and parking information and data. By analyzing this data, facility management and indoor positioning systems can optimize the performance and utilization of the office space. The aim for the customer is to decrease operational cost by selecting the right workplace strategy to increase the productivity of the employees. Sensors are placed in the chair and show whether it is occupied or available for use. The location sensors enable navigation through the building in order to find an available chair and desk.

Company Name: Silent Space
Website: http://www.silent-space.fr/en
Company Name: Orosound
Website: http://www.orosound.com/
Target user: Office users, tenants, facility managers, and property managers
What they do: The new trend of shared office space and open office space brings new problems, including increased noise level. This problem is addressed by two French companies in a different way. One strategy is to install a sensor-based noise diffuser in the office so private conversations cannot be heard by others. Another way of using a shared office and staying concentrated is using noise cancelling and filtering earphones.

Better communication

Company Name: Equiem
Website: https://www.getequiem.com/
Target user: Building owners and property managers
What they do: In order to understand the benefits and problems of a building, the owner needs to be able to communicate with the tenant, and the occupier should have access to all information about the building. This company provides a technology- and cloud-based information and communication platform app, where not only can tenants and landlords communicate with each other much faster and more transparently, but also different tenants of the same building can post information about activities and events in and around the building. Direct communication can also be a great way to act on emergency incidents immediately.

Company Name: Allthings
Website: https://www.allthings.me/en
Target user: Building owners and property managers
What they do: This company is a communication platform which enables the owner, the asset manager, or the property manager to communicate directly and digitally with tenants. With this, the user of the building is not only able to receive information about the building in real time, but can also report any incidents in the building via this digital platform. The building owner can decide to add additional offers for the community, such as a concierge service, access to restaurants or bars in or around the building, laundry services, and many more services for everyday needs.

Digital community

Product Name: Tishman Speyer's Zo app
Website: https://findyourzo.com/
Product Name: Cushman & Wakefield's Workplace Edge
Website:
https://appadvice.com/app/workplaceedge/1272122320
Company Name: Office App
Website: http://www.getofficeapp.com/
Target user: Building owners and property managers
What they do: Companies who are leasing buildings from these providers can offer these apps to their employees with the aim of increased work-life balance. The app lets the employee's book amenities like a yoga class, catering services, a haircut, and other daily needs. At the end, the building will not only be a place to work but also an entire service provider for all the daily needs. These firms provide also a technology- and cloud-based information and communication platform app, where tenants and landlords can communicate with each other much faster and more transparently; also, different tenants of the same building can post information about activities and events in and around the building.

Company Name: Workwell
Website: http://www.workwell.io/
Target user: Building owners and property managers
What they do: This company provides a digital platform for the building to perform tasks like booking meeting rooms. It also adds a community platform for the tenant's employees, who can access the services in and around the office building. It is able to integrate all of the building systems into the platform, for example, access to the cafeteria, infrastructure, or room controls. The users of the platform can control all this via an app. In the end, the aim is to increase productivity, the wellbeing of the employees, improved

utilization of the services within the building, and offer the company a more connected workplace.

Company Name: WeLive
Website: https://www.welive.com/
Company Name: Ollie
Website: http://www.ollie.co/
Company Name: Common
Website: https://www.common.com/
Company Name: The Collective
Website: https://www.thecollective.com/
Company Name: Mindspace
Website: https://www.mindspace.me/
Company Name: Mofang Gongyu
Website: http://www.52mf.com.cn/
Company Name: OneThird
Website: https://www.urbancampus.com/
Target user: Building owners and property managers
What they do: A focus of these co-living space providers will be to provide a digital platform where the tenants can interact with each other. Hence, a building is a community platform. The tenants are connected and can communicate with each other until they meet for events or sports together.

Heating Ventilation Air Conditioning (HVAC) solutions

Company Name: PHYSEE
Website: http://www.physee.eu/
Target user: Owners, asset managers, and facility managers
What they do: This company from the Netherlands has developed a fully transparent smart window, which also produces electricity. Using data and information about the exterior and interior of the building, the smart window is able to automatically control ventilation and the use of sun blinds. Technology enables the reduction of energy costs and increases quality of air and daylight within the building.

Company Name: Enerbrain
Website: https://www.enerbrain.com/
Target user: Facility managers
What they do: This company uses sensors to obtain data about temperature, CO_2, and humidity in buildings to analyze energy efficiency and the effectiveness of the building's HVAC system. With an algorithm, the firm assesses the data using machine learning to improve efficiency over time. The dashboard displays the optimized results, which are visible on a computer or smart device. The firm is claiming that they have reduced energy bills by one third.

Company Name: uHoo

Website: https://uhooair.com/

Target user: Office users, tenants, facility managers, and property managers

What they do: This company from Hong Kong has created an indoor air sensor that can detect nine different air quality parameters. The data can be tracked and analyzed digitally through the Internet or an app. For commercial properties, the software can be interoperable with other building management systems.

Company Name: MeetFlo

Website: https://meetflo.com/

Target user: Contractors, facility managers, and property managers

What they do: This company provides a water leak detection system with an automated shut-off mechanism. Leakage of water can be detected in a pipe with the help of a sensor in the valve. This smart device can find this leakage even if it is hidden.

Analytics for elevators

Company Name: Dynamic Components GmbH
Website: https://www.dynamic-components.de/en/
Target user: Contractors, facility managers, and property managers
What they do: One of the major problems leading to high maintenance and repair costs is the inability to know what is happening until a component in your building breaks. This can be avoided by using sensors and transmitting performance data from the component to see its health status. This firm collects data from elevators and analyzes it to provide sufficient information to make a sophisticated decision on whether or not to act, and to repair elevators before they break down entirely.

Company Name: Lift Technology
Website: http://lift-technology.de/
Company Name: Uptime
Website: https://www.uptime.ac/
Target user: Contractors, facility managers, and property managers
What they do: These companies from Germany and France focus on maintaining the running time of elevators and reducing the maintenance costs by using sensors. These sensors transmit data that can be quickly analyzed to determine whether repairs are required. This can reduce planned and unplanned downtimes. It reduces expenditures, as maintenance requirements can be predicted, and repairs can be conducted in advance. The data can be used to transparently compare productivity and cost of different facilities.

Interoperability

Company Name: Areo
Website: http://areo.io/
Target user: Contractors, facility managers, and property managers
What they do: This Norwegian company provides a digital interoperability platform for facility managers and real estate stakeholders. Facility managers can optimize daily operations and can provide better insight about the building to the property owner. He can also keep track of all operations within the building and approve repairs and maintenance requirements for a contractor to execute. After the completion of the work, the contractor can hand over required documents via the digital platform. This digital platform provides some model management, document management, forms and process management, floor plan management, and has maps and collaboration features to eventually analyze the data.

Company Name: bGrid
Website: http://bgridsolutions.com/
Target user: Owners, tenants, and property managers
What they do: This company from the Netherlands provides a one-stop smart building solution, which connects people with the building through their smart phones to control the climate, blinds, light, and even the coffee machine. The company provides an API platform with third-party hardware and software developers to ensure interoperability with their smart building applications. Hence, one of the main aims of the company is to ensure that the different devices are communicating with each other.

Company Name: Thing Technologies GmbH
Website: http://www.thing-it.com/
Target user: Facility managers
What they do: A smart building becomes intelligent by enabling communication through interoperability of different devices, technologies, sensors, and gateways. The company provides a digital solution and a platform for computer-aided facility management, which integrates services and software systems to automate and digitize business processes and task management after receiving the information from the building. For example, the building, which is connecting devices, data, and services, might identify a malfunction and request work to be completed by users such as service technicians, janitors, the cargo team, or a doctor. These systems can be provided for smart offices, smart hotels, smart hospitals, smart farming, or even smart energy supply. The firm is currently working on the Cube in Berlin, which might become the smartest building in Europe.

Smart key solutions

Company Name: August Home - Assa Abloy
Website: http://august.com/
Target user: Owners, tenants, and property managers
What they do: This San Francisco-based smart lock producer creates digital and technology-focused door locks. These are very handy for short-term rentals or to give access to a certain group of people only. Technology enables them to use the smart locks also via a smart phone and to open or close the door from a different location. This firm has been acquired by the large Swedish lock producer Assa Abloy.

Company Name: Nuki
Website: https://nuki.io/en/
Company Name: igloohome
Website: https://www.igloohome.co/
Target user: Owners, tenants, and property managers
What they do: These companies from Austria and Singapore also provide a smart key and smart access solution aiming to replace the physical key with a digital platform that can be controlled via an app.

Company Name: Uniberry GmbH
Website: https://cido.io/de/
Target user: Owners, tenants, property managers, and the logistics industry
What they do: This company provides a smart access solution focusing on the logistics industry, which requires a solution for redelivery problems. It provides on-demand access to the building management system for deliveries if the receiver is not at home. This digital solution enables the delivery to be scanned at the front door of the building to gain access in order to leave the postal packets at a predetermined location within the building.

Smart homes

Company Name: Homey
Website: https://www.athom.com/en/homey/
Target user: Residential property owners and tenants
What they do: This company from the Netherlands is focusing on the residential market and connects different smart home appliances so they can be controlled with only one device and app. They ensure interoperability with an open platform that connects more than eight different wireless technologies and hundreds of brands.

Company Name: Nest
Website: https://nest.com
Target user: Residential property owners and tenants
What they do: Acquired by Google, this company serving the private and residential real estate market provides digital thermostats, cameras, doorbells, and smoke detectors. These devices are interconnected and make the home of the user smart. Users are able to use their devices via an app and control their home from a different geographic location. Furthermore, it is possible to control heating and cooling of the property to optimize energy consumption.

Company Name: Sense
Website: https://sense.com/
Company Name: Verv
Website: https://verv.energy/
Target user: Residential property owners and tenants
What they do: This company is focused solely on the optimization of energy consumption within a private residential home. The device collects data after it is installed in the property's electrical panel to analyze energy use and provide a solution to reduce energy consumption. As it is a digital platform, it is accessible via the web and app.

Company Name: Geomatrix
Website: https://geomatrix-retail.com/
Target user: Retail space renters
What they do: Before conducting a marketing or product campaign in a different geographical location, it is important to analyze information about the market and consumer behavior. This company provides a digital solution for predictive analytics within the retail sector. It aims to increase the impact of retail space and marketing campaigns. The user should be able to select the most suitable replication by analyzing data such as road and traffic speed, points of interest, active population on social networks, consumer behavior, distribution, active population using mobile phones, et cetera. Hence, a company that is looking for new retail space can outsource data creation by using the solution of this company.

Rent collection

Company Name: Acasa
Website: https://www.helloacasa.com/
Target user: Landlords
What they do: It can be a daunting task for landlords to chase payments for utility bills. This company has created an app which helps the processes of moving in, managing utility bill payments, and creating transparency. A mobile app helps to see which payments have been completed, which are outstanding, and upcoming. This technology-based software can also help to reduce workload for the landlord.

Company Name: Zenhomes GmbH
Website: https://www.vermietet.de/
Target user: Landlords
What they do: This company from Germany tracks rental income directly from the bank account and processes notices, if arrears appear. Furthermore, it analyzes data and provides recommendations for improvements.

Rent guarantee

Company Name: Rentify
Website: https://www.rentify.com/
Target user: Landlords
What they do: This company guarantees rental income for the landlord through a software platform without the hassle of dealing with tenants. It utilizes technology and data to analyze and value the property in order to determine the exact market price for the rent. The company makes a profit by renting out the property to the tenants for a higher price than the rental payment to the landlord.

Company Name: Rented.com
Website: https://www.rented.com/
Target user: Landlords and property managers
What they do: This company brings the landlord and the property manager together. The aim of the owner of the property is often to have full occupancy, without any arrears. The property manager would like to offer a rent guarantee, but he does not want to take such a risk. This firm uses big data and technology to enable the property manager to provide a rent guarantee and to earn its commissions for services to the landlord. Backed by a large fund, this firm is actually providing an insurance policy to the landlord and the property manager.

Digital contracts

Company Name: LeBonBail
Website: https://www.lebonbail.fr/
Target user: Landlords
What they do: This company enables a landlord to write and sign a contract digitally with a tenant. It has the huge advantage that the landlord does not need to be close to his property geographically and can still let out his premises

Smart hotels

Company Name: Hostmaker
Website: https://hostmaker.com/
Target user: Landlords
What they do: This company understood the emerging trends in the hospitality and real estate sectors and has provided a solution providing synergy between those industries. The trend of short-term lettings of residential properties that are more like smart hotels with concierge are a target market for this platform. Landlords can pass on the management of the property to this technology-driven company. The company then takes on the entire process of marketing and renting out the flat to the short-term tenant.

Company Name: Wunderflats
Website: https://wunderflats.de/
Company Name: Homelike
Website: https://www.thehomelike.com/
Target user: Landlords
What they do: Business travelers might need to stay more than a month in a location, but do not always want to stay in a hotel, preferring something in between a hotel and an apartment. A furnished apartment that can be rented on a long-term or mid-term basis is often the ideal solution. These two companies from Germany provide a platform similar to Airbnb that focuses on business travelers who are staying a minimum of one month in the flat. The flats are fully furnished and provide a similar standard as a hotel. They also provide cooking facilities very often. Because of this, this hybrid model of an apartment and hotel is predicted to be a strong growing market.

Maintenance

Company Name: Roomhero
Website: https://www.roomhero.de/
Target user: Landlords and asset managers
What they do: This company has reacted to the emergence of the furnished apartment market by providing a fully digital platform where owners of apartments can use the service in order to furnish their properties. They are an integrated interior designer and decorator. The platform provides a one-stop solution where landlords can change the premises to business apartments, co-living spaces, or even student apartments. The entire process from conception to delivery and assembly of the furniture will be provided.

Company Name: Thermondo
Website: https://www.thermondo.de
Target user: Landlords and asset managers
What they do: The transformation to the digital world can also mean a simple step of having a website that provides more transparency. This company from Germany has established a digital solution for optimizing the processes of installing radiators and thermostats. All information is given on the website, so clients are able to purchase their services and products directly online. With this strategy, the firm became the largest installer of heating systems in Germany.

Company Name: Doozer Real Estate Systems GmbH
Website: https://dasist.doozer.de/
Company Name: Travaux.com
Website: https://www.travaux.com/
Company Name: 123devis.com
Website: https://www.123devis.com/
Company Name: MyHammer AG

Website: https://www.my-hammer.de/
Company Name: MyBuilder Limited
Website: https://www.mybuilder.com/
Company Name: HomeStars, Inc.
Website: https://homestars.com/
Company Name: IKEA Group
Website: http://taskrabbit.co.uk/
Target user: Homeowners, tenants, property managers, contractors, and craftsmen
What they do: These marketplaces connect homeowners, tenants, or property managers with contractors and craftsmen, a very helpful tool to fix damage or defects in a building rapidly. These marketplaces are actually an equivalent to the Yellow Pages, but are digital, faster, and more transparent. Contractors must collect references and display the work they have completed in the past. This way, the client can judge the performance and select the most suitable offer. The task of finding a contractor, which might have taken a few weeks in the past, can be completed in a few days, or even hours.

Company Name: SMS Assist
Website: https://www.smsassist.com/
Target user: Asset managers and property managers
What they do: This company provides a multi-site property maintenance platform for clients who own a portfolio of properties. It has automated and standardized repair and maintenance requests and works. With accurate data, it is able to provide the user with an in-depth analysis of the performance of their individual properties. The contractors are able to use their mobile devices to monitor, communicate, and report maintenance and reparation work completed. The client is able to decrease costs in the long run due to the efficiency and the standardization of the processes.

6

Construction

"The whole difference between construction and creation is exactly this: that a thing constructed can only be loved after it is constructed; but a thing created is loved before it exists."

–Charles Dickens

The construction industry provides a good bridge between the rapid movement in the previous categories to things still evolving into the future—a lot of the construction tech, from 3D design (and printing), to adding sensors throughout new home and buildings, to modular construction—the challenge is to integrate all these things with systems that communicate with each other, and we are not a long way away from that in homes and larger construction.

BIM & ConTech

The construction industry is a huge market, where small companies and large companies are coming together to construct a building or infrastructure. According to the Construction Intelligence Centre, the size of the global construction industry is going to grow to US$10 trillion by 2020.

Independent of the size and scope of the construction project, there are always a number of individuals, small companies, and large companies participating in the design and construction of the building. The stakeholders in such a project are, among others, the investor or the owner, who is hiring a general construction management company, who then hires an architect, engineer, and general contractor, and they are then hiring the tradespeople or subcontractors. Outside of the circle, there are also the suppliers and the local government councils, who are involved and have a say in the construction project. So, such a project can be very complex and overwhelming, which in the past required a lot of manual reporting and controlling.

In this chapter, we are going to look at the way buildings have been constructed in the past, and what disadvantages and challenges the industry had to face. We are analyzing what is going to happen if the industry does not change in the future.

Afterwards, we will analyze the technology that was not available in the past, but has evolved over the last decade, and that can now be implemented for construction and real estate. We are looking at Building Information Modelling (BIM) and how it can reduce delays and work that has to be redone at the construction site, and increase transparency due to sophisticated planning.

We are defining Construction Technology (ConTech) as the implementation of technology in the construction industry and focusing on the improvements of off-site construction. Here, we are looking at different examples where components or even entire buildings are constructed in factories and assembled at the construction site.

Constructing the old way

The old way of constructing a building was using the design, bid, and build method, which means the owner hires an architect to design the building. After the design has been drawn, the general contractor will be hired, who then hires the subcontractors to execute the construction work. This method was very problematic, as the input of the general contractor, engineers, and subcontractor was not integrated in the design process.

So the integrated project delivery method, where the architect, general construction manager and project manager worked together, has become standard practice. This is a method where all the stakeholders work together very closely. Nevertheless, in this method, the construction manager is at risk to deliver on time and within budget. So his aim is to implement the best possible ways to improve the situation so that he can really deliver on time and within budget.

In the past, when stakeholders were using old-fashioned printed paper blueprints at construction sites, the controller had to be always on site. When they saw problems at the site, they had to run back to the office and communicate and change everything there. Hence, he had to re-do a lot of work, which resulted in higher costs and deadlines not being met. Smart and innovative start-ups also have found solutions to use technology to increase safety, e.g. by tracking the location and movement of all people at the construction site.

The future of construction

The construction industry of the future is going to provide the conventional way of developing the building, but use digital tools at the construction site, and install prefabricated buildings. In order to reduce the problems of speed, transparency, and cost, the construction industry is focusing heavily on planning. Digitally planning and coordinating before execution are the most important aspects of the new way of construction. New firms in the Construction Technology (ConTech) industry are providing digital controlling and management platforms.

Furthermore, sensors are changing conventional brick and mortar assets to smart buildings, which use building information modelling software to digitally visualize, access, manage and maintain the property. These systems also go well beyond this aspect, with software and communications designed to create true "just-in-time" logistical support to cut into downtime on site.

Building Information Modelling (BIM)

BIM is the digital representation of a building, and it describes the processes before, during, and after the construction of a building. Through the digitization, and with compatibility, it enables stakeholders to visualize a project, and communicate. It enables architects to communicate their visual designs and ideas to engineers, who have requirements on sustainability, structure, and mechanical, electrical, and plumbing (MEP). Contractors can see a simulation of the building and estimate the cost before constructing. Finally, facility managers will be in a position to use all this information digitized in one place to efficiently maintain the building. This can be a number of processes or components, which can be interlinked.

Nevertheless, not all systems and software are compatible with each other. Hence, the start-up BIMwelt Systems GmbH from Germany is focusing on providing advice to companies who are obliged to use BIM for projects where it is mandatory. Certain government projects will require the implementation and utilization of BIM within their project. The responsibility of a contractor is to be able to show that he is BIM-ready by being able to detail the process he is utilizing to adapt to the project's standards and requirements set by the government. In order to avoid the BIM compliance trap, BIMwelt Systems GmbH is working together with architects, specialist planners, component manufacturers, general contractors, and other participants in the construction project to provide innovative solutions, as not all BIM software solutions work with each other.

Similarly, the UK BIM Alliance is an organization aiming toward the digital formation of the Built Environment. This means that all stakeholders and participants of over 2.1M people, 95% of which work in SMEs (Small- and Medium-Sized Enterprises) in this industry, are supposed to use BIM Level 2 as business as usual by 2020.

As there is no way back, we are going to compare a number of different software solutions for different categories, which will be used in the future before, during, and after construction. We are differentiating between software for architecture, sustainability, structures, mechanical, electrical, and plumbing (MEP), construction simulation, estimation and analysis, and facility management.

The architectural design and documentation software applications that are widely used are Autodesk Revit Architecture, Graphisoft ArchiCAD, Nemetschek Allplan Architecture, Gehry Technologies - Digital Project Designer, Nemetschek Vectorworks Architect, Bentley Architecture, 4MSA IDEA Architectural Design (IntelliCAD), CADSoft Envisioneer, Softtech Spirit, aRhinoBIM (BETA), etc.

These programs provide computer-aided solutions for handling all common areas of aesthetics and engineering throughout the entire planning and design phase of the construction project, which includes the building, interior, and urban areas.

Building performance analysis tools allow designers to simulate the sustainability of the built environment. Widely used are Autodesk Ecotect Analysis, Autodesk Green Building Studio, Graphisoft EcoDesigner, IES Solutions Virtual Environment VE-Pro, Bentley Tas Simulator, Bentley Hevacomp, DesignBuilder, etc. They enable the designer, architect, and engineers to calculate the sustainability performance within the context of the building model. This helps to optimize energy efficiency and to reduce the carbon footprint to comply with building energy requirements.

These software solutions also automatically outline suitable bioclimatic architecture strategies for a project, while seeking opportunities to keep costs appropriate. Finally, the aim is to generate documentation and reports that are compliant with accreditations like the LEED certification. The aim for all software should be interoperability, which means the ability to import building layouts directly from different file formats to smoothly reuse existing data without the requirement for third-party applications to interpret data.

Structural analysis software efficiently creates a model of the design and building for the structural engineer to analyze aspects like reinforced concrete, post-tensioned concrete, steel, wood, cold-formed steel, aluminum, and masonry. The commonly used software are Autodesk Revit Structure, Bentley Structural Modeler, Bentley RAM, STAAD and ProSteel, Tekla Structures, CypeCAD, Graitec Advance Design, StructureSoft Metal Wood Framer, Nemetschek Scia, 4MSA Strad and Steel, Autodesk Robot Structural Analysis, etc.

Structural engineering firms will use a tool to guarantee maximum reliability of the building structure and be able to frame entire projects, while simultaneously creating schedules, material cut lists, framing elevations, and fully dimensioned 2D shop drawings for different national standards.

The MEP engineer will use the software to visualize high-level details of the components he is going to install in the building and support the coordination with other contractors and stakeholders. Commonly used software solutions for MEP engineers are Autodesk Revit MEP, Bentley Hevacomp Mechanical Designer, 4MSA FineHVAC + FineLIFT + FineELEC + FineSANI, Gehry Technologies - Digital Project MEP Systems Routing, and CADMEP (CADduct / CADmech). Engineers are able to draw building models in 3D using fabrication-ready components and objects such as HVAC services and systems in order to calculate the module within the BIM model. This digital pre-planning decreases the number of errors during the construction on-site.

A construction planner can use simulation software to review and analyze the construction digitally to estimate time and cost. The planner will be able to edit and manage a construction project throughout its lifecycle using the visual context of 4D, which includes temporary works, logistics, costs, and resources, with the aim of improving cost, quality, health, and safety before, during, and after the construction. A single visual interface will enable 4D time simulation, photorealistic rendering, and PDF-like publishing.

The commonly used software solutions are Autodesk Navisworks, Solibri Model Checker, Vico Office Suite, Vela Field BIM, Bentley ConstrucSim, Tekla BIMSight, Glue (by Horizontal Systems), Synchro Professional, and Innovaya, which provides quality assurance solution for BIM validation, compliance control, design review, analysis, and code checking. They aim to improve coordination, communication, produce optimized construction schedules, maximize efficiency and meet the distinctive needs of the various construction process trades and phases.

After the completion of the construction, the digital data that has been collected during the construction via BIM systems has to be transferred and handed over to the facility manager. This can maximize the return on the asset and facility with solutions for the entire building lifecycle, as it provides visibility into construction, and ensures design and construction information is retained for operational use. For example, when the contractor is required to replace filters, pipes, and other operational maintenance, he can access 3D models to identify the exact location, size, model and construction year of the component. This avoids a lot of guessing.

Among others, these are the software solutions for the facility managers: Bentley Facilities, FM: Systems, FM: Interact, Vintocon ArchiFM (For ArchiCAD), Onuma System, and EcoDomus.

Construction Technology (ConTech)

Digital off-site construction

High efficiency and effectiveness can be achieved through the use of technology and optimization in the planning, designing, fabricating, assembling, and production processes. This also is used for the prefabrication or modular construction of the building. Even though it is not a new concept, off-site construction became very popular in recent years. Due to increased land and construction prices, developers had to find a way to enhance the value chain, decrease cost, and optimize production.

In the context of the construction technology industry, we are differentiating between prefabrication and modular construction, which are both part of off-site construction. Prefabrication means that some parts of a building will be built in a factory and transported to the construction site for final assembly.

Modular construction means that the prefabricated components are built as boxlike models. Hence, it is a type of prefabrication. Firms like Alphabet, CitizenM, Muji's, and Marriott have already started to buy and invest in modular and prefabricated buildings.

Furthermore, prefabricated shipping containers are showing a solution for the increased price of construction. These shipping containers are remodeled in a factory into small offices or residential properties, which can then be delivered to any location in the world. The production is cheaper, faster, and provides a state-of-the-art interior design, while fulfilling all legal requirements. In this fast-changing world, where technology determines lifestyle, buildings built a hundred years ago cannot fulfill today's technological requirements and can be replaced easily with those containers on a regular basis.

Innovative Firms

On-site document access and convocation

Company Name: PlanGrid
Website: https://www.plangrid.com/
Target user: Developers, contractors, surveyors, and more
What they do: This company understood the difficulty of taking floor plans, blueprints, documents, and other paperwork to the construction site. Therefore, with the invention of the iPad and other mobile devices, it enabled contractors and their workers to see these paper documents digitally on mobile devices. Furthermore, the stakeholders are now able to view, annotate, and communicate the information directly via this mobile device. This means, for example, that users can see a defect at the construction site and take a picture of it with a mobile device and upload it directly with comments to the cloud-based database. This cloud-based document sharing and reporting system seems to be a very simple tool, but it is very powerful, as the current standard at a lot of construction sites in the world is still the use of emails and paper blueprints.

Company Name: PlanRadar
Website: https://www.planradar.com/
Target user: Surveyors, developers, and contractors
What they do: This company based in Austria is providing a simple, but very powerful technological solution that helps a number of stakeholders within the real estate and construction industry to digitally identify and monitor defects. The process is very simple: it is possible to upload a floor plan or a blueprint of the property into a cloud-based system and conduct an inspection of the building. On a smart phone or an iPad, it is possible to see the floor plan and your location and to mark the location of the defects. This can be particularly very helpful for surveyors who have to determine technical details of the building.

Company Name: Fieldwire
Website: https://www.fieldwire.com/
Target user: Developers, contractors, surveyors, and more
What they do: This company also enables workers on a construction site to be more flexible. They're able to use blueprints, floor plans, tasks managers, checklists, pictures, specifications, documents, and instructions in the office and at the construction site to collaborate online and offline. The 360-degree photos are efficient and provide up-to-date information from the construction site to participants not at the site.

Company Name: APROPLAN
Website: https://www.aproplan.com/
Target user: Investors, architects, designers, engineers, and contractors
What they do: This company provides an app that enables stakeholders such as the investors, architects, designers, engineers, and contractors to communicate through a centralized platform where all the drawings and documents of the projects can be found. This means a request for change from an investor can be seen by the other stakeholders, who can report back immediately on the cost of such a change or even the plausibility of execution. As a cloud-based and digital platform, it replaces any paper documents and can be used on site.

Company Name: Honest Buildings
Website: https://www.honestbuildings.com/
Target user: Project controllers, investors, banks, or owners
What they do: A number of stakeholders, like investors, banks, or owners will not be at the construction site. However, it is crucial for them to track the progress of the project, the capital that has been spent already, and further capital requirements. It is quite costly to hire a controller, who is also not able to be on site all the time. Hence, the use of project management or controlling

software that gives up-to-date information about the development, orders from suppliers, status of delivery of those components, and other relevant information is crucial to minimize cost. It does not mean a controller is not going to be hired, but it is good for an investor to understand the status of his project. This software provides a solution for those participants who cannot be on site, and reduces the work for the controller, who does need to be on-site all the time. It is a leading data-driven project management and procurement platform for the commercial property market. It includes a reporting function, so private equity real estate investment firms, project managers, and operators can communicate with each other.

Company Name: Assignar
Website: https://www.assignar.com/
Company Name: ProperGate
Website: http://propergate.co/
Company Name: Sablono
Website: https://www.sablono.com/en/
Target user: Developers, contractors, and suppliers
What they do: On a construction site, a number of activities are happening at the same time and it can be very difficult to oversee operations. These three companies provide a cloud-based software that enables contractors to transparently track in real time the location of the workforce, assets, and materials, and status of compliance. Thus, it improves the operation's productivity through collection of data, workforce management, timely planning, scheduling, and communicating tasks. This full-field mobility solution software focuses on the speed of decision making.

Company Name: Baufolio+
Website: http://baufolio.de/
Target user: Developers, contractors, surveyors, controllers, and more
What they do: The founders of this company, who have legal backgrounds, understood that important decisions are made verbally or through a short email, which can have a significant impact on the outcome of the construction, and thus legal liabilities. Hence, they created an app that complies to legal requirements to sign off on work such as acceptance, objections, and detected defects at the construction site. It supports agreements on deadlines, costs, and wages, which can be signed off directly at the construction site by all parties, without the need to go back to the office.

Company Name: Built
Website: http://www.getbuilt.com/
Target user: Construction loan administrators, banks, and lenders
What they do: This company is the provider of a cloud-based construction lending software that transparently connects the investor and the borrower and reduces risk, improving the borrower experience. By simplifying compliance, it is able to increase loan profitability for small, regional, and national lenders.

Data collection and analytics

Company Name: Uptake
Website: https://www.uptake.com/
Target user: Equipment managers, project managers and contractors
What they do: This company collects data at the construction site and from machinery in order to analyze and predict different scenarios, such as the breakdown of a machine. It tells when a machine is going to break down when it is being used at low, medium, and full capacity. Hence, the engineer can determine when to increase or decrease the speed. It can be used to avoid delays, and increase productivity, cyber security, reliability, and safety. The interesting part is that it is using machine learning and data science by analyzing the information.

Company Name: Archilyse
Website: https://www.archilyse.com/
Target user: Property investors, architects, and more
What they do: This company is providing a software-as-a-service solution, which analyzes property with the help of a floor plan and gives information based on pure math and science. It closes the discrepancy between design, architecture, geometry, and math. The aim is to make design and architectural qualities more objective and comparable. This means that the company provides a software that analyzes the floor to determine the value of the entire property.

Company Name: Spacemaker AI
Website: https://www.spacemaker.ai/
Target user: Architects, real estate developers, and municipalities
What they do: The company's aim is to provide software and a technological solution enabling design of a building that is efficient and effective for the desired usage, but also provides the best and the highest quality in terms of living space or working space. The software calculates the best possible solutions to deliver a building with all required attributes and characteristics. Artificial intelligence in the cloud-based software enables it to provide statistics and the best possible solutions. Architects, real estate developers, and municipalities can use the software to design a building by entering parameters in order to receive the ideal building.

Company Name: BuildSafe
Website: https://www.buildsafe.se/en/
Target user: Developers and contractors
What they do: Health and safety is certainly one of the most important aspects during the construction project. This company provides a mobile tool to create a safe and secure construction environment.

Building materials

Company Name: Celitement
Website: http://www.celitement.de/en/
Company Name: Solidia Technologies
Website: http://solidiatech.com/
Company Name: Smart Crusher B.V.
Website: https://www.slimbreker.nl/smartcrusher.html
Target user: Developers and contractors
What they do: These three firms are also aiming to reduce the environmental impact during the production of construction material. They have their patented methods of producing concrete, cements and stone. Celitement focuses on energy efficiency with hydraulic binders by also reducing the input of raw materials and carbon dioxide emissions. Solidia, on the other hand, focuses on the reduction of water consumption (60%–80%) and carbon footprint (by up to 70%), while producing concrete. Instead of water, they use CO_2, using manufacturers' existing infrastructure, specifications, formulations, raw materials, and production methods. Finally, the Smart Crusher uses concrete waste to produce climate-neutral new concrete.

Company Name: Woodoo
Website: http://woodoo.fr/home/
Target user: Developers and contractors
What they do: By re-engineering certain characteristics like fire-retardance, weatherproofing, and sturdiness, this firm enables wood to become an important material in the construction industry again. It is reinventing this material to be used during and after construction to solve environmental challenges such as climate change through their reduction of carbon emissions with their zero-net-carbon production.

Company Name: Materialize.X
Website: http://materializex.com
Target user: Developers and contractors
What they do: This company uses data science combined with chemistry to reengineer wood. It delivers an optimized manufacturing process by using state-of-the-art machine learning optimization algorithms to create sustainable materials instead of toxic adhesive.

Company Name: Imerys
Website: http://imerys.com/
Target user: Developers and contractors
What they do: The construction industry not only makes improvements in the digital implementation of information, but also uses new materials to provide more productive, sustainable, efficient, and effective materials. This company provides specialty solutions out of minerals that provide specialty functions, for example to improve coating, coverage, mechanical strength, and heat resistance.

Company Name: Wellsun
Website: https://www.wellsun.nl/
Target user: Developers and contractors
What they do: The company provides a full glass facade solution that not only replaces conventional walls, but also generates more energy than the building consumes. It uses an algorithm to selectively shield direct and intense light by also turning it into electricity. Hence, it is responsible for glare and heating of the building to facilitate an ideal indoor climate.

On-site automation and robotics

Company Name: Fastbrick Robotics
Website: https://www.fbr.com.au/
Company Name: Construction Robotics
Website: http://www.construction-robotics.com/
Company Name: CyBe
Website: https://cybe.eu/
Target user: Developers and contractors
What they do: These firms have created bricklaying machines that are 3D robotic systems. This can enhance waste management, safety, and accuracy. An example is the increase in productivity of bricklaying work at the construction site.

Company Name: Kewazo
Website: https://www.kewazo.com/
Target user: Developers and contractors
What they do: This German company has found the solution to reduce cost by up to 30% with their automated system for scaffolding installation. It also decreases the number of workers needed.

Prefabricated buildings

Company Name: Ten Fold Engineering
Website: https://www.tenfoldengineering.com/
Target user: Investors, house buyers, and end consumers
What they do: This company brings efficiency to the next level by providing a building that automatically unfolds. They can be delivered anywhere in the world, ready to use within a few minutes of arriving. The container-like building can also be folded back together and shipped to a different place.

Company Name: Containerwerk eins GmbH
Website: https://www.containerwerk.com/
Target user: Investors, house buyers, and end consumers
What they do: This company from Germany is taking old shipping containers and refurbishing them into living or working space.

Company Name: Tempohousing
Website: http://www.tempohousing.com/
Target user: Investors, house buyers, and end consumers
What they do: This company from Europe is also recycling shipping containers and changing them into affordable and modern residential living space. They use automated serial processes to prefabricate containers to build more economical, efficient, and sustainable accommodations.

Modular construction

Company Name: Katerra
Website: https://katerra.com/
Target user: Architects, developers, and contractors
What they do: This firm is optimizing material planning, manufacturing, supply chain management during the design, construction, and development process of the building project. The technology enables production within a factory of prefabricated components, which will then be delivered on time to the site with the overall aim of reducing delay in time and unnecessary costs.

Company Name: FullStack Modular (FSM)
Website: http://www.fullstackmodular.com/
Target user: Architects, developers, contractors, and investors
What they do: This company has built the tallest prefabricated modular building, with 32 floors, by using the design-to-build process and incorporated Building Information Modelling (BIM). It focuses on multi-family buildings, hotels, and student housing.

Company Name: WASP
Website: http://www.wasproject.it/w/en/
Target user: Investors, house builders, and end consumers
What they do: This company focuses on manufacturing 3D printers for off-site construction usage. The aim is to provide a sustainable way to construct entire buildings. WASP (World's Advanced Saving Project) is a self-claimed open-source project from Italy focusing on developing 3D printing. The aim is to provide the knowledge for the end consumer to 3D print a building.

Company Name: Toyota Housing
Website: http://toyotahousing-id.com/en/
Company Name: Muji
Website: https://www.muji.com/jp/mujihut/en.html
Target user: Investors, house buyers, and end consumers
What they do: Japan has a sophisticated prefabricated real estate market, which has been constructing buildings and factories since the 1960s. These two firms are focusing on quality and sustainability, while building efficient residential houses. Toyota's houses have a 60-year guarantee and can be assembled within 45 days.

7

The Future

"Don't listen to what they say.
Go see."

–Chinese Proverb

The blockchain technology

After looking at so many different start-ups that are changing and improving the real estate industry by making it faster, more transparent, and reducing the cost, we can see that their main aim was to use the Internet and a given technology to improve the status quo.

However, following the financial crisis in 2008, a few people have probably questioned the financial structure and the system which led to this crisis. They have invented the blockchain, which is in my opinion the logical next step for the Internet. It is new and has a different mechanism and actually can be used for all sorts of business solutions. One of the main points is that it is decentralized and open, which can democratize access through a blockchain solution to everyone.

Information now is centralized, stored in large data centers, and accessible to the owner only. This means, for example, information about a transaction is centrally located within the bank that stores it in a data center.

With the blockchain on the other hand, information is decentralized and saved at different devices in different parts of the world. This information is immutable or unchangeable, because if a hacker is trying to change it, he has to change it on all systems at the same time and has to change the history of the information, which is saved on the blockchain simultaneously.

Nevertheless, it is in its early stage and it can be quite difficult to understand this technology. It was also difficult to understand the possibilities of the Internet in the 1990s. Because of this, there are a number of people and institutions who do not believe that the blockchain will ever change their industry, nor provide solutions which are more efficient, effective, transparent, cheaper, and faster.

The blockchain provides a number of functions and forms and cannot be simplified for one usage only. It can be described as a decentralized and distributed database, which is recording transactions in a secured ledger system. The data or information is saved chronologically and stored in this chain. It is decentralized and does not give control to only one individual or institution. Instead it is saved on a number of computers, which makes the information immutable.

Furthermore, it enables a transaction to be authenticated and approved through the system in real time without the usage of a middleman, such as a broker, a bank, a notary, or a government institution. The advantage of implementing blockchain within the real estate industry is a faster transaction, full transparency, and the reduction of cost.

Blockchain solutions for the real estate industry

A number of people and institutions believe in blockchain technology, and show it by investing, experimenting, and utilizing the technology for their business purposes.

The blockchain is a technology that will penetrate not only the residential property market, but also the commercial real estate industry. Cryptocurrencies, such as Bitcoin or the alternative "Altcoins," will overcome boundaries so that the local real estate market will not have any geographical limitations, and make real estate accessible to everyone globally.

Faster transaction

By using the blockchain, it is possible to transfer cryptocurrencies within seconds or minutes from one wallet to another. There is not any difference if a property buyer from London sends money from his crypto currency wallet to a seller located in London, Tokyo, or New York; they only have to determine which currency to use. There are no geographical boundaries. Furthermore, the transaction can be authorized by someone who has been approved. This person can use digital authentication through a blockchain. After all, there is no intermediary who is controlling the transaction; instead it is a peer-to-peer transaction that will be recorded in the decentralized blockchain.

Full transparency

As every transaction will be saved in a blockchain, it is easy to see the full history. Everyone is able to access this information after receiving authentication. The information and data are also immutable, which means it is not possible to change the information at one computer, as all the other computers will have their origin information. This decentralization makes it possible to access this information from everywhere in the world.

Smart contracts will enable the contract management process to become faster, smoother, and more transparent. All participants will work on fully automated sales and lease agreements, where even the payments are determined and fully automated so that the funds will be transferred automatically, even outside of working hours, if the predetermined conditions are met. The blockchain technology is able to verify identities and make background checks of participants who are able to use the personal digital key to gain access in order to authorize a smart contract or transaction. This could arguably reduce the probability of fraud.

Cost reduction

The transaction runs through this blockchain system, which is decentralized and requires computerized authentication without human intervention. As the transaction does not require any middlemen, the cost of completing a transaction is the cost of computing power. This cost can be optimized, automatized, and reduced. Furthermore, all information about the transaction is saved in the blockchain and cannot be amended, which could make such a transaction even more safe.

Not only start-ups, but also large institutions such as banks or governments are thinking of finding solutions to their current problems through blockchains. Within the real estate context, governments are partnering with start-ups who can use blockchains to digitize the land registry and facilitate any new land registry records to the blockchain system.

Another example is the mortgage application process at a bank. They understand that their mortgage application process is not the fastest, and their systems are not fully standardized and automatized. Hence, they are exploring new ways to speed up the identification of the mortgagee, process the due diligence, and make the approval faster.

Firms like Airbnb have disrupted an entire industry. Start-ups working with blockchains are questioning the high fees that Airbnb takes from landlords and guests and are trying to provide a more cost-efficient solution through the use of blockchains, which could then disrupt Airbnb in turn.

Property prices have increased steadily, and property transactions are a long-term process, which makes property difficult to afford and illiquid. These circumstances do not correspond to the lifestyle of a millennial who might not want to settle in one location until the mortgage expires. They want to be flexible geographically, but also benefit from real estate investment opportunities, or even sell their property after a short period.

Blockchain start-ups are creating real estate exchanges, where it is possible to buy a fraction or a share of the property or even a development project. It is then possible to sell or trade these shares globally. An example could be that a small investor from Africa invests in a property in Canada with only US$100, and benefits from the price appreciation of the property.

A number of governments are enabling blockchain to secure land registries with the support of start-ups. The following states aim to implement the blockchain technology to secure and track property titles: Ghana, Georgia, Ukraine, Sweden, Estonia, Rwanda, Brazil, Dubai, and the Indian state of Andhra Pradesh. This could lead to changes on how real estate is bought, sold, leased, and evaluated in the rather distant future.

Innovative Firms

Blockchain technology

Company Name: Bee Token
Website: https://nest.beetoken.com/
Target user: Residential property owners leasing out properties
What they do: This company provides a similar service to Airbnb, but it is decentralized and uses blockchain technology. Once landlords decide to be paid via a cryptocurrency, the service is much cheaper than the one from the main competitor, as the automated smart contract is offering a tokenized short-term or long-term rental without the approval of a middleman. The payment will be triggered automatically and received in the firm's cryptocurrency.

Company Name: Rentberry
Website: https://rentberry.com/
Target user: Residential property owners and private tenants
What they do: This firm is focusing on the rental market to allow landlords to find tenants and tenants to find rental properties. It has standardized and automated the rental process. Tenants can submit their personal information, landlords can receive the credit report and make custom offers, where both parties can digitally sign a rental agreement and use the online rental payment system. It is now focusing on using the Ethereum blockchain technology in order to decentralize smart contracts signed by the landlord and the tenant. The aim of the platform is to eliminate the broker and his fees through this automation.

Company Name: Atlant
Website: https://atlant.io/
Target user: Investors, landlords, and tenants who want to use real estate exchanges
What they do: This company provides a tokenized ownership of a property and also gives the option to globally rent out a property via a peer-to-peer system. By tokenizing the ownership of the property, the vendor is able to list his real estate where tokens will be created to represent the share of the property. At the decentralized blockchain exchange platform, it is possible to sell, trade, and buy tokens directly from the counterpart. The peer-to-peer decentralized rental platform also enables the rental of a property globally and directly without an intermediary. As this is a fully automated process and does not require any middleman or broker, the fees can be kept low.

Company Name: Token Estate
Website: https://tokenestate.io/
Target user: Asset Managers and investors who want to use real estate exchanges
What they do: This company is an efficient real estate fund marketplace, where asset managers are able to tokenize their funds and investors are able to invest in real estate funds through blockchains. Real estate managers are able to present their funds to the company in order to raise money. The company completes a thorough due diligence of the asset manager and once approved, tokens for the fund will be issued and listed at the company's exchange platform. Individual or institutional investors are able to invest in real estate funds globally through the exchange platform provided by this firm.

Company Name: Etherty
Website: https://etherty.com/
Company Name: Trust-X
Website: https://trustx.io/
Target user: Investors who want to use real estate exchanges
What they do: These companies from Dubai and Prague are creating real estate exchanges. Real estate vendors will be able to list their properties at the exchange, in order to create shares (tokens). The investors are able to buy the property with the platform's cryptocurrency. Furthermore, it is possible to search for properties globally to buy, trade, and sell those shares and tokens via the platform.

Company Name: BitRent
Website: https://bitrent.io/
Target user: Investors
What they do: This company creates investment opportunities into the early phase of a construction project for the average investor. It is a construction company, which was founded in 1995 and recently entered the crypto market. Investors are able to buy tokens, which will be used to fund the construction project. Using Building Information Modelling (BIM) and Radio-frequency identification (RFID), the company creates transparency during the construction process, where the investor is able to track the status of the progress digitally. By entering a smart contract, the buyer will become a shareholder in the property and receive a return or can trade or sell the property.

Company Name: Caviar
Website: https://caviar.io/
Target user: Investors
What they do: The high volatility of the crypto market can be a large problem for investors. This company aims to diversify the investments by providing not only investments in crypto tokens, but also by raising funds in order to invest in real estate, too. During market depreciation, investments into real estate through debt are supposed to create negative correlation to the crypto market. Hence, the investment strategy will be more balanced.

Company Name: Bitland
Website: http://www.bitland.world/
Target user: All governments and local citizens (of Ghana), who want to use a blockchain land registry
What they do: Ghana is using blockchain technology to create an automated land registration process and a digital system of recording that is decentralized and difficult to manipulate. It aims to provide facilities so that locals and organizations are able to register their properties and land record data directly on the blockchain.

Company Name: ChromaWay
Website: https://chromaway.com/
Company Name: Midasium
Website: http://midasium.herokuapp.com/
Target user: Governments, banks, and financial institutions who want to use a blockchain land registry
What they do: These two companies are aiming to make the real estate ecosystem more transparent and faster by using blockchain. Land titles, mortgage registrations, and tenancy agreements will be saved in a global decentralized ledger. Through smart contracts, it is possible to record transactions and create a permanent history of the real estate. Those smart contracts will be

a replacement for tenancy agreements, lease contracts, and mortgage agreements. This can speed up the entire transaction process as smart contracts do not require days or months to settle like traditional paper contracts.

Company Name: Tellus Title Company
Website: https://www.tellustitle.com/
Target user: Governments
What they do: This San Francisco-based start-up is using the blockchain technology in order to transfer and record real estate titles. They have created a real estate blockchain, which is using a geo-coded blockchain protocol concept that is patented and uses a universally integrated property identifier system. The system will record the rights of property owners using the blockchain network that is open and decentralized.

Company Name: Propy
Website: https://propy.com/
Target user: Governments, investors, and sellers
What they do: This company was executing on the distributed public blockchain network Ethereum the world's first blockchain-based real estate transaction. By using the Ethereum blockchain network, it has used a smart contract to settle the transfer of ownership where the purchase price was paid with its own token PRO (Propy). The purchase price was the equivalent of $60,000. This made it also the first ever real estate cryptocurrency transaction in the Ukraine. The firm has partnered with the Ukraine government which confirmed to pilot the blockchain title registration.

Company Name: Ubitquity
Website: https://www.ubitquity.io/
Target user: All governments and citizens of Brazil
What they do: This company provides a software-as-a-service blockchain platform. This enables tracking and recording property registers and title ownership, and the uploading of record documents into the blockchain. It is in the process of integrating the registry information of the Land Records Bureau in Brazil onto the blockchain platform. Furthermore, this company was the first one creating a platform on blockchain that was securing real estate record keeping.

Company Name: Imbrex
Website: https://imbrex.io/
Target user: Investors, landlords, brokers, and private and public institutions using data analytics
What they do: This company from the United States provides a decentralized global real estate data marketplace. It uses the Ethereum blockchain, where data is locally sourced, validated, and saved through smart contracts and crypto economics. Investors, landlords, brokers, and firms are able to access and curate the decentralized database. Local data was not accessible globally as it was geographically isolated and centralized. The aim of this marketplace is to collect and connect data in order to make it accessible worldwide to all stakeholders.

Company Name: LegalThings One
Website: https://legalthings.io/
Target user: Property managers, owners, and tenants who want to use smart contracts
What they do: This company focuses on creating contracts that are a hybrid between a paper contract and the blockchain-based smart contract. This means a contract is digitized, decentralized, and saved as a contract on the blockchain, which is an immutable audit trail. This could mean a property manager digitally writes the contract with the tenant, who also agrees with its digital signature to the contract. The firm's aim is to automate the contract lifecycle via its platform.

8

Conclusion

"A tree starts with a seed."

–Indian proverb

The growth of PropTech start-ups

The real estate industry is more stable than other markets with strong foundations, and everyone is well connected. It has a well-functioning ecosystem, which is very well-funded. By entering such an environment, it is certainly not easy for a start-up to change things around. Disrupting such an ecosystem might require to inventing something new that is as stable as a brick and mortar building. Investors might not want to be exposed to the volatility of the cryptocurrency market and want to remain with their stable real estate investment.

Start-ups entering this market with the intention of disrupting the entire real estate industry will face a lot of rejections. Having said that, it is at the same time obvious that the real estate market has to change, has to adapt to new technology, and has to increase efficiency and effectiveness in order to tackle the problem of real estate scarcity.

Start-ups with the intention of creating value through better transparency, increased speed, and lower cost for the existing processes will receive a lot of support from the real estate industry. It is therefore important for those smart young entrepreneurs to create a solution which makes an immediate impact. They have to take baby steps by introducing small improvements. Instead of thinking about artificial intelligence or blockchains, it could be enough to create a simple digital solution at the beginning to get the foot in the door.

This doesn't mean that the artificial intelligence and the blockchain solution is not required in the near future. However, providing such a solution involving a big change now to the real estate market might not feel tangible for those property players, who might not even be able to realize what potential is available already.

Nevertheless, the number of start-ups within the property environment has increased, and the venture capital investments appreciated, too. This trend will sustain in the near future.

The future of the PropTech industry

PropTech start-ups have increased because they have realized the trend within the real estate industry. There is a scarcity of available spaces in gateway cities, and a shift in thinking happened. The best asset managers will implement value add and opportunistic strategies for their portfolios and invest in gateway cities and logistics properties.

On the other hand, retail properties are closing, as users shift to online retailers, and tenants are using their properties for a shorter period. Hence, there are a number of providers who are offering small furnished apartments for business travelers with flexible conditions. WeWork is also offering a very flexible contract to work, as there are more freelancers now than 50 years ago. There is also less human intervention for the entire process, from reserving a place to checking it out. This is also possible because buildings are becoming smart, and the city provides enough information for someone from a different location to find all necessary information and locations. Someone who enters a new location is able to utilize its devices in order to enter a new community within a building or a city within seconds.

Becoming smart also means the introduction of new competitors from a different industry. Smart homes and sensor devices are not produced by real estate companies. Instead, IT firms are able to deliver such products and obtain the valuable data from the building and its tenants.

If the property industry is not cooperating with start-ups, there will be other industries that will enter the real estate field. Firms like Google or Amazon are already providing products for the real estate market. They will utilize the existing infrastructure—a building of steel, or brick and mortar—and utilize it for their benefits.

It has to be the aim of the real estate industry to cooperate with start-ups who want to create additional value for the property, the processes, and the real estate ecosystem. Those firms adapting new technology will lead the way in obtaining data and assets. Firms that are not adapting to new technology will still possess a valuable asset, but they are not making the most out of it.

The real estate industry might have leapfrogged a number of digital inventions and did not adapt quickly to the digital changes. However, with smart buildings able to communicate and the blockchain, the real estate industry might have found its door to the 21st century. If the real estate industry is opening up to the blockchain and uses smart contracts in a smart city or a smart building, it could become the biggest competitor to Google, Amazon, and Apple.

It is now the time for the smart and brave who want to use technology. They will be able to enter the market and gain fast market share, because it is not money which makes them rich. Instead, it is the ability to create value and attractive dividends for the investors.

Further Information

For further information please visit:

www.PropTechGuide.com

References & Footnotes

References

Baum, A. (2018). PropTech 3.0: The Future of Real Estate. [online] Oxford: University of Oxford. Available at: https://www.sbs.ox.ac.uk/school/news/property-tech-30-%E2%80%93-ground-breaking-report-looks-future-real-estatert%2FPropTech%25203%2520-%2520The%2520Future%2520of%2520Real%2520Estate.pdf&usg=AOvVaw01P3kMpHdUbb1doscxeTrj [Accessed 1 Jun. 2018].

Rivaton, R. and Pavanello, V. (2018). Make Real Estate Great Again: Proptech, Real Estech, Construction Tech.

Mougayar, W. and Buterin, V. (2016). The Business Blockchain. Wiley.

Ross, A. (2017). The Industries of the Future. London: Simon & Schuster.

Brynjolfsson, E. and McAfee, A. (2016). The Second Machine Age. Norton & Company.

Antonopoulos, A., Hariry, S., Lords, M., Morgan, P., Scothorn, M. and Zolt-Gilburne, S. (2017). The Internet of Money. Merkle Bloom LLC.

Lifthrasir, R. (2018). 10 Reasons ICOs Harm Blockchain Real Estate, and 2 Better Alternatives. [online] Available at: https://medium.com/@RagnarLifthrasir/10-reasons-icos-

harm-blockchain-real-estate-and-2-better-alternatives-
90e3be997f38 [Accessed 1 Jun. 2018].

The Investor. (2018). How Blockchain can Impact Real Estate
Investment. [online] Available at:
https://www.theinvestor.jll/news/world/others/blockchain-
reshaping-real-estate-industry/ [Accessed 1 Jun. 2018].

Deloitte United States. (2018). Blockchain in Commercial Real
Estate (CRE) | Deloitte US. [online] Available at:
https://www.deloitte.com/us/en/pages/financial-
services/articles/blockchain-in-commercial-real-estate.html#
[Accessed 1 Jun. 2018].

Bravenewcoin. (2018). 0x Price Analysis - Liquidity Plagues the
DEX » Brave New Coin. [online] Available at:
https://bravenewcoin.com/news/0x-price-analysis-liquidity-
plagues-the-dex/ [Accessed 1 Jun. 2018].

Castro Betancourt, S. (2018). Real Estate Bitcoins Are Here:
This Is How Blockchain will Transform Real Estate. [online]
Property Portal Watch. Available at:
http://www.propertyportalwatch.com/real-estate-bitcoins-
arrive-blockchain-will-transform-real-estate/ [Accessed 1 Jun.
2018].

Dale, B. (2018). Real Estate ICOs Are Moving In, But Investors
Aren't Floored - CoinDesk. [online] CoinDesk. Available at:
https://www.coindesk.com/real-estate-icos-moving-investors-
arent-floored/ [Accessed 1 Jun. 2018].

Bridge, C. and Soo, J. (2018). Blockchain in Residential &
Commercial Real Estate | Investment Bank. [online]
Investmentbank.com. Available at:

https://investmentbank.com/blockchain-real-estate-transactions/ [Accessed 1 Jun. 2018].

Frumkin, D. and More, R. (2018). The Rise of Asset-Backed Cryptocurrencies - Invest In Blockchain. [online] Invest In Blockchain. Available at: https://www.investinblockchain.com/asset-backed-cryptocurrencies/ [Accessed 1 Jun. 2018].

Firms	Website	Page
123devis.com	https://www.123devis.com/	132
21st Real Estate	https://www.21re.de/en/	30
Acasa	https://www.helloacasa.com/	128
Allthings	https://www.allthings.me/en	117
ALLVR GmbH	https://allvr.net/	76
APROPLAN	https://www.aproplan.com/	148
Archilyse	https://www.archilyse.com/	151
Areo	http://areo.io/	123
Assetti	https://assetti.pro/	107
Assignar	https://www.assignar.com/	149
Atlant	https://atlant.io/	167
August Home - Assa Abloy	http://august.com/	125
Baufolio+	http://baufolio.de/	150
Bee Token	https://nest.beetoken.com/	166
Better Mortgage	https://better.com/	61
bGrid	http://bgridsolutions.com/	123
Bitland	http://www.bitland.world/	169
BitRent	https://bitrent.io/	168
Blend	https://blend.com/	58
Breather	https://breather.com/	33
Brickowner	https://brickowner.com/	52
BrickVest	https://brickvest.com/	46
BuildSafe	https://www.buildsafe.se/en/	152
Built	http://www.getbuilt.com/	150
Cadre	https://cadre.com/	47
Canopy	https://findyourcanopy.com/	55
CapitalRise	https://www.capitalrise.com/	51
Caviar	https://caviar.io/	169
Celitement	http://www.celitement.de/en/	153

ChromaWay	https://chromaway.com/	169
CoAssets	https://www.coassets.com/	47
Common	https://www.common.com/	119
Commuty	https://www.commuty.net/	34
CompStak	https://compstak.com/	81
Construction Robotics	http://www.construction-robotics.com/	155
Containerwerk eins GmbH	https://www.containerwerk.com/	156
CrediFi	https://www.credifi.com/	56
Credit Sesame	https://www.creditsesame.com/	54
Cushman & Wakefield's Workplace Edge	https://appadvice.com/app/workplaceedge/1272122320	118
CyBe	https://cybe.eu/	155
DataScience Service	https://ds-s.at/	82
Deepki	https://www.deepki.com/	110
Disruptive Technologies Research AS	https://www.disruptive-technologies.com	112
Distech Controls	http://www.distech-controls.com/	111
Doozer Real Estate Systems GmbH	https://dasist.doozer.de/	132
Dronotec	http://www.dronotec.com/	78
Dynamic Components GmbH	https://www.dynamic-components.de/en/	122
EMonitor	http://emonitor.ch/	109
Enerbrain	https://www.enerbrain.com/	120
Enlighted	http://www.enlightedinc.com/	112
Entic	http://www.entic.com/	111
Envelope	https://envelope.city/	31
Equiem	https://www.getequiem.com/	117
Equifax	https://www.equifax.co.uk/	54
EstateGuru	https://estateguru.co/	61

Etherty	https://etherty.com/	168
EVANA	https://www.evana.de/	73
Exporo	https://exporo.de/	49
Exquance	https://exquance.com/	107
EyeSpy360	http://www.eyespy360.com/	74
FairFleet360	https://www.fairfleet360.com/	77
Fastbrick Robotics	https://www.fbr.com.au/	155
Fieldwire	https://www.fieldwire.com/	148
Finastra	https://www.finastra.com/	57
FirstFuel	http://www.firstfuel.com/	110
FullStack Modular (FSM)	http://www.fullstackmodular.com/	157
Geomatrix	https://geomatrix-retail.com/	127
Geospin GmbH	http://www.geospin.de/en/	80
Giraffe360	http://giraffe360.com/	74
Go-PopUp	http://www.gopopup.com/	34
Greenbird Vertriebs Gmbh	https://cleanbird.fm/	113
Guaranteed Rate	https://www.guaranteedrate.com/	61
Habiteo	https://pro.habiteo.com/	74
Habito	https://www.habito.com/	59
HABX	https://www.habx.fr/	80
Homelike	https://www.thehomelike.com/	131
HomeStars, Inc.	https://homestars.com/	133
Homey	https://www.athom.com/en/homey/	126
Honest Buildings	https://www.honestbuildings.com/	148
Hostmaker	https://hostmaker.com/	131
HouseCanary	https://www.housecanary.com	73
Housers	https://www.housers.com/	50
Houzeo	https://www.houzeo.com/	79
Hypcloud	https://www.hypcloud.de/en/	59

ICapital Network	http://www.icapitalnetwork.com/	45
iDeals Solutions Group	https://www.idealsvdr.com/	72
igloohome	https://www.igloohome.co/	125
IKEA Group	http://taskrabbit.co.uk/	133
Imbrex	https://imbrex.io/	171
Imerys	http://imerys.com/	154
Insider Navigation	http://insidernavigation.com/	75
Intralinks	https://www.intralinks.com	72
Katerra	https://katerra.com/	157
Kewazo	https://www.kewazo.com/	155
Knock	https://www.knock.com/	29
Kofax	https://www.kofax.com/	57
KOMA Ltd.	http://www.koma.guru/	74
LandInsight	https://landinsight.io/	31
LeBonBail	https://www.lebonbail.fr/	130
LegalThings One	https://legalthings.io/	172
LendingHome	https://www.lendinghome.com/	62
LendingTree	https://www.lendingtree.com/	59
Leverton	https://www.leverton.ai/	73
Lift Technology	http://lift-technology.de/	122
Locatee	https://www.locatee.ch/en/	114
Lone Rooftop	http://lonerooftop.com/	114
Mashvisor	https://www.mashvisor.com/	82
Materialize.X	http://materializex.com	154
Matterport	https://matterport.com/	74
MeetFlo	https://meetflo.com/	121
Midasium	http://midasium.herokuapp.com/	169
Mindspace	https://www.mindspace.me/	119
Moderan	http://www.moderansolutions.com	106

Mofang Gongyu	http://www.52mf.com.cn/	119
Muji	https://www.muji.com/jp/mujihut/en.html	158
MyBuilder Limited	https://www.mybuilder.com/	133
MyHammer AG	https://www.my-hammer.de/	132
MyNotary	https://www.mynotary.fr/	83
NavVis	https://www.navvis.com/	75
Nest	https://nest.com	126
Nested	https://nested.com/	28
Nuki	https://nuki.io/en/	125
Nutonian	https://www.nutonian.com/	110
Office App	http://www.getofficeapp.com/	118
Ollie	http://www.ollie.co/	119
OneThird	https://www.urbancampus.com/	119
Opendoor	https://www.opendoor.com/	28
Orosound	http://www.orosound.com/	116
PHYSEE	http://www.physee.eu/	120
Placense	https://www.placense.com/	111
PlanGrid	https://www.plangrid.com/	147
PlanRadar	https://www.planradar.com/	147
Point	https://point.com/	53
Popertee	http://popertee.com/	33
Prodigy Network	https://www.prodigynetwork.com/	47
ProperGate	http://propergate.co/	149
Property Moose	https://propertymoose.co.uk/	52
Property Partner	https://www.propertypartner.co/	51
Propy	https://propy.com/	170
Purplebricks	https://www.purplebricks.co.uk/	27
Quicken Loans	https://www.quickenloans.com/	62
Rate Reset	http://www.ratereset.com/	58

Real Capital Analytics	https://www.rcanalytics.com/	81
RealAtom	https://realatom.com/	60
Reali	https://www.reali.com/	27
realiz3d	http://realiz3d.fr/	76
RealMassive	https://www.realmassive.com/	81
RealPage	https://www.realpage.com/	108
Realty Mogul	https://www.realtymogul.com/	46
RealtyShares	https://www.realtyshares.com/	48
RealX.pro	https://realx.pro/	30
REalyse	https://realyse.com/	82
Redfin	https://redfin.com/	27
Rentberry	https://rentberry.com/	166
Rented.com	https://www.rented.com/	129
Rentify	https://www.rentify.com/	129
Roomhero	https://www.roomhero.de/	132
Roostify	https://www.roostify.com	58
Sablono	https://www.sablono.com/en/	149
Sense	https://sense.com/	127
Sensorberg GmbH	https://www.sensorberg.com/	115
ShareDnC	http://www.sharednc.com/	31
Shojin	https://www.shojin.co.uk/	50
Silent Space	http://www.silent-space.fr/en	116
Sindeo	https://www.sindeo.com/	62
SkenarioLabs	https://skenariolabs.com/	79
Skycatch	https://www.skycatch.com/	77
Smart Crusher B.V.	https://www.slimbreker.nl/smartcrusher.html	153
SMS Assist	https://www.smsassist.com/	133
SoFi	https://www.sofi.com/	63
Solen	https://www.solen.co/	80

Solidia Technologies	http://solidiatech.com/	153
Source Central	https://www.sourcecentral.co/	45
Spacebase	https://www.spacebase.com/	33
Spacemaker AI	https://www.spacemaker.ai/	152
Spaceti	https://spaceti.com/	115
Spectando	http://spectando.com/	75
Spotscale	http://spotscale.com/	77
Storage Share	http://www.storage-share.nl/	32
StoreFront	https://www.thestorefront.com/	33
Stowga	https://www.stowga.com/	32
StrideUp	https://www.strideup.co/	53
Tellus Title Company	https://www.tellustitle.com/	170
Tempohousing	http://www.tempohousing.com/	156
Ten Fold Engineering	https://www.tenfoldengineering.com/	156
The Collective	https://www.thecollective.com/	119
The eLocations	http://www.elocations.com/	32
Thermondo	https://www.thermondo.de	132
Thing Technologies GmbH	http://www.thing-it.com/	124
Tishman Speyer's Zo app	https://findyourzo.com/	118
Token Estate	https://tokenestate.io/	167
Toposens	https://toposens.com/smart-building/	114
TOWER360 GmbH	http://www.tower360.co/	108
Toyota Housing	http://toyotahousing-id.com/en/	158
Travaux.com	https://www.travaux.com/	132
Triplemint	https://www.triplemint.com/	26
Trussle	https://trussle.com/	60
Trust-X	https://trustx.io/	168
Ubitquity	https://www.ubitquity.io/	171

uDA. urban Data Analytics	http://www.urbandataanalytics.com/	81
uHoo	https://uhooair.com/	121
Uniberry GmbH	https://cido.io/de/	125
Unmortgage	https://capital.unmortgage.com/	53
Uptake	https://www.uptake.com/	151
Uptime	https://www.uptime.ac/	122
Verv	https://verv.energy/	127
VISN	https://www.visn.io/en	113
VTS	http://www.vts.com/	106
Walliance	https://www.walliance.eu/	49
WASP	http://www.wasproject.it/w/en/	157
WeLab	https://www.welab.co/en	63
WeLive	https://www.welive.com/	119
Wellsun	https://www.wellsun.nl/	154
WiredScore	https://wiredscore.com/	111
Woodoo	http://woodoo.fr/home/	153
Workwell	http://www.workwell.io/	118
Wunderflats	https://wunderflats.de/	131
Yanport	https://www.yanport.com	26
Yodlee	https://www.yodlee.com/	55
Yopa	https://www.yopa.co.uk/	27
Zenhomes GmbH	https://www.vermietet.de/	128
Ziroom	http://www.ziroom.com/	55

CPSIA information can be obtained
at www.ICGtesting.com
Printed in the USA
LVHW052127150520
655578LV00004B/90